A CANDLELIGHT INTRIGUE

CANDLELIGHT ROMANCES

214 Pandora's Box, BETTY HALE HYATT
215 Safe Harbor, JENNIFER BLAIR
216 A Gift of Violets, JANETTE RADCLIFFE
217 Stranger on the Beach, ARLENE HALE
218 To Last a Lifetime, JENNIFER BLAIR
219 The Way to the Heart, GAIL EVERETT
220 Nurse in Residence, ARLENE HALE
221 The Raven Sisters, DOROTHY MACK
222 When Dreams Come True, ARLENE HALE
223 When Summer Ends, GAIL EVERETT
224 Love's Surprise, GAIL EVERETT
225 The Substitute Bride, DOROTHY MACK
226 The Hungry Heart, ARLENE HALE
227 A Heart Too Proud, LAURA LONDON
228 Love Is the Answer, LOUISE BERGSTROM
229 Love's Untold Secret, BETTY HALE HYATT
230 Tender Longings, BARBARA LYNN
231 Hold Me Forever, MELISSA BLAKELY
232 The Captive Bride, LUCY PHILLIPS STEWART
233 Unexpected Holiday, LIBBY MANSFIELD
234 One Love Forever, MEREDITH BABEAUX BRUCKER
235 Forbidden Yearnings, CANDICE ARKHAM
236 Precious Moments, SUZANNE ROBERTS
237 Surrender to Love, HELEN NUELLE
238 The Heart Has Reasons, STEPHANIE KINCAID
239 The Dancing Doll, JANET LOUISE ROBERTS
240 My Lady Mischief, JANET LOUISE ROBERTS
241 The Heart's Horizons, ANNE SHORE
242 A Lifetime to Love, ANN DABNEY
243 Whispers of the Heart, ANNE SHORE
244 Love's Own Dream, JACQUELINE HACSI
245 La Casa Dorada, JANET LOUISE ROBERTS
246 The Golden Thistle, JANET LOUISE ROBERTS
247 The First Waltz, JANET LOUISE ROBERTS
248 The Cardross Luck, JANET LOUISE ROBERTS
249 The Searching Heart, SUZANNE ROBERTS
250 The Lady Rothschild, SAMANTHA LESTER
251 Bride of Chance, LUCY PHILLIPS STEWART
252 A Tropical Affair, ELISABETH BERESFORD
253 The Impossible Ward, DOROTHY MACK
254 Victim of Love, LEE CANADAY
255 The Bad Baron's Daughter, LAURA LONDON
256 Winter Kisses, Summer Love, ANNE SHORE

THE GOLDEN LURE

Jean Davison

A CANDLELIGHT INTRIGUE

Published by
Dell Publishing Co., Inc.
1 Dag Hammarskjold Plaza
New York, New York 10017

This book was first published under the title
The Golden Torrent.

For information write: Doubleday & Company, Inc.,
245 Park Avenue, New York, N.Y. 10017

Dell ® TM 681510, Dell Publishing Co., Inc.

ISBN: 0-440-12965-6
Printed in the United States of America
Reprinted by arrangement with
Doubleday & Company, Inc.

First Dell printing—May 1979

PROLOGUE

The man who called himself Carlos Ortega turned the dark-skinned beach vendor aside with a curt shake of his head and waggle of his forefinger. A green tourist might have relied on a spoken "no," but Ortega knew the procedure the natives used to discourage a sales pitch, and he had followed it to the letter. The dark man with the hammocks shrugged philosophically and moved off down the beach.

After a few more paces toward the water Ortega unfolded the wooden chair he'd been carrying under his arm and sat down. In a moment he opened the Mexico City newspaper he had bought downtown and began to read, glancing up now and then to be sure no one else was about to approach him. This was Ortega's favorite time of day, the morning hours before the tourists poured out of the high-rise luxury hotels and cluttered up the beaches. It was only eight-thirty, quite early by Acapulco's standards, and the long crescent of glittering white sand was almost deserted.

Not far away from him a young woman was watching a child play in the flat, shallow waves. Further along a solitary swimmer bobbed in the water and a middle-aged Mexican fisherman was rowing toward shore. Down the beach in the other direction two men were unraveling long ropes tied to a red and white parachute spread out at their feet. Later that morning they would attach the ropes to a motorboat and begin offering parasail rides around the bay to anyone brave enough to try it. The hammock vendor stopped to talk with them about something. Ortega watched the three men suspiciously. It occurred to him that they might be discussing him. When they continued talking without glancing his way, he went back to his newspaper.

Ortega hated people. He had always hated people, although he told himself that his aversion to crowds was only a sensible precaution he had been forced to take. His wife Margarita used to chide him about it. She would pout, like all women, *You never take me anywhere.* She would say, *Don't be a ninny. Who could possibly recognize you at a movie theater in Buenos Aires?*

But stranger things had happened. Hadn't one of his old colleagues come face to face with a Jewess he had known in 1944 on a streetcar in Montevideo? Twenty years had gone by and yet she recognized him at once and began to scream and tear at his clothes. He had escaped only by jumping off the car while it was still moving. And Rudolf Schwandt was a man one might look at every day and not really see. He was a *Dutzendgesicht*, a dime-a-dozen face.

Ortega liked to think that his own face was not so easily forgotten. In his youth he had had the classic features that appeared on countless political posters: high forehead, strong cheekbones, sharp blue eyes. Although he had lacked both a good education and important family connections, his appearance and bearing had carried him a long way.

Even now that his hair was gray and his shoulders drooped a little despite his conscious efforts, he had seen younger women look at him. . . . But he hadn't been interested in anyone since Margarita died. Margarita, his beloved wife. He always took his winter vacation at Acapulco because they had honeymooned there.

He went through the paper in his usual methodical way. First the front page, then the editorials. The lead editorial was on the latest terrorist outrage, another political kidnapping. He savored every word. Eventually, he was certain, some editor in the West would be compelled to say, *Maybe the Nazis were right*. Or, more timidly perhaps, *Hitler may have been wrong about many things but he was right when he talked about the need for a new World Order. That need has never been greater than it is today*. . . . But it would be a long, long time before anyone had even that much courage. He knew he might not live to see it, but he was certain the day would come.

He skimmed the entertainment section, noting in passing an advertisement for yet another anti-Nazi movie. Such things no longer irritated him. He had finally come to realize that the Jews who controlled the American movie industry would keep harping on the same theme until they dropped. They simply could not help themselves. They were eaten up with hate and resentment.

And he had to admit they had been extraordinarily successful in turning the swastika into an unattractive symbol. Hitler too had known the uses of propaganda. Simply keep stressing the same theme over and over and soon everyone will believe it.

Without glancing up Ortega reached into his shirt pocket and pulled out a piece of hard butterscotch candy. It was a habit he had picked up after he quit smoking. He twirled open the noisy cellophane, rolled the ball around in his mouth, and cracked it open with his strong molars.

He was about to put the paper aside when a small headline on one of the back pages caught his eye. *Nuevo Hallazgo Arqueologica en Austria.* It was the mention of Austria that interested him, not the fact that there had been a new archaeological discovery. But it was the dateline that truly riveted his attention. The single word TELLIN seemed to jump off the page. He stared at the word for a full minute, and all the old memories came back.

Tellin had been his last outpost during the war, a fallback position during the lengthy retreat of spring 1945. Not once since then had he ever seen the name in print anywhere. Tellin was a village in the Tirolean Alps. He and his men had stopped there only because the old castle overlooking the village was like a makeshift fortress. It gave them a chance to rest in a secure place.

He read the newspaper article, still mentally translating the Spanish into German despite all the years of using the other language. The article said a fisherman on the lake at Tellin had tangled his nets on a group of submerged wooden pilings. These pilings, it turned out, were extremely old. They had once supported a walkway leading to a Bronze Age lake dwelling, a group of "twig-and-mud" houses that had been raised over the lake. The article said that the archaeological community was elated because the discovery offered an opportunity to examine the remnants of a lake dwelling that were still underwater and had therefore never been disturbed by souvenir hunters, as were most of those found previously in dry lake beds in Switzerland and Germany.

Ortega was extremely interested in the Tellin lake.

He skipped the other paragraph of archaeological twaddle and went to the last sentence: "Professor Albert T. Windle of the Archaeology Department at Rossalyn College in Maryland, an expert in underwater archaeology, says that he plans to lead an excavation of the Tellin site this summer if funding can be obtained and if a qualified crew of young volunteer divers can be assembled."

Ortega's heartbeat quickened in excitement and hope.

In 1945 his name was Heinrich Krueger. He wore the twin lightning bolts on his collar, he was a *Schutzstaffel Obersturmführer*, a high-ranking SS officer. By the time his exhausted troops reached Tellin, the U.S. Seventh Army under Patch was closing in from the north to join the French in a pincers movement from the northwest. To the south the U.S. Fifth Army was advancing toward Austria out of Italy. Every day the noose tightened. Because he was a realist, Krueger had already made his plans for escape.

There was only one last thing to take care of.

As leader of the last SS contingent leaving Innsbruck it had fallen to him to take charge of a part of the German treasury that had been sent there earlier for safekeeping. The four steel boxes containing a small fortune in gold ingots were too heavy to take out of the country with him. They would have to be left behind. Somewhere.

The command in Berlin had been of no help whatsoever. Several months before, he had noticed how the chain of authority had quietly begun to disintegrate. In the first years of the war Berlin had kept up a drumfire of specific orders—*do this, do that*—and they would question the slightest deviation . . . but by the end of 1944 there were fewer directives. Those that came were nonspecific, almost lackadaisical. For the first time one of them had actually read, "Use your own judgment in this matter." That single sentence had convinced him that the war was lost.

During those last days in Tellin he thought of the future. In a few days, a week at most, whatever was left of the German government would capitulate. Then the Allies would begin to fight it out among themselves—the Americans and the British against Russia, with Russia probably being the loser after several more years of devastation. In the end all Europe would be an ash heap. He wasn't able to imagine what might follow that. He didn't care what followed that.

It didn't occur to him to try to plan for a day when he might be able to come back to Austria as a civilian and get the Reich's gold for himself. He was above all a soldier. Imaginary orders formed in his mind.

This treasure must not fall into enemy hands.

But how? He gave it a good deal of thought.

Once the missing gold was traced to Tellin, as he assumed it might be, the Americans would search the castle thoroughly, probably digging up the grounds. He considered having his men row the boxes out to the middle of the lake and toss them overboard, but rejected this solution as too obvious. The American frogmen would go over every inch of the lake bed. He could see the triumphant smile on the American general's face when a subordinate brought him word that the gold had been found. No, he wouldn't let them outwit him so easily.

But for a long time he couldn't think of a better way. Then one afternoon as he was destroying papers in his quarters, it came to him.

Earlier, immediately after arriving at the castle, he had ordered a routine reconnaissance of the surrounding area. During a careful search of the mountain spur on which the castle stood, a corporal had looked behind some brush and discovered a rough hole in the rock a little over a meter in diameter. Being both thorough and inquisitive, the young soldier had crawled inside and found a bare cave about the size of two large rooms. This seemingly trivial detail had been included in a lengthy written report.

After burning the last of his papers, Krueger decided to inspect the cave himself. The entrance was widened with pick and shovel to allow him to walk inside. The passage was long and narrow; by the time he reached the cave proper the sleeves of his field jacket were smudged with dirt.

The interior was quite dark. The junior officer who had walked in ahead of him directed his flashlight all over the inside. The floor and walls were smooth and wet in an unpleasant way. Krueger remembered feeling as though he had stepped inside the stomach of some gigantic beast. He was in a hurry to leave.

The cave floor sloped downhill toward the rear, where there was a pool of water about two and a half meters across. He walked to the edge of it and noted that it looked deep. An underground spring, he thought. The water was a peculiar shade of blue, as though it were throwing off a dim light of its own. There were faintly bluish reflections on the nearby wall of rock. He didn't think much about this at the time, but many years later he would think about it a great deal.

After midnight the boxes were brought down to the cave from the castle wine cellar. The entrance was sealed with an explosive charge. To the sleeping village the muffled sound would be indistinguishable from the soft plumping of the distant artillery that had been going on for several days now. Standing outside, Krueger examined the spot where the cave entrance had been. The opening had disappeared into the surrounding rubble of loose stone.

It was a cool night, he remembered, and there was a nearly full moon. Looking around him, he could barely make out a few of Tellin's darkened houses. His eyes swept to his right across the broad lake and off toward the black mountains where the American and French armies waited.

He was pleased. No one would ever "liberate" this part of the Reich's treasury. There was only a pang of regret that no one in Berlin would ever commend him for this job well done. He turned to one of the men and said, "*Saubere Arbeit.*" Nice work. He was speaking as much to himself as to them. He glanced up at the castle tower's silhouette and began the hard climb back to his quarters.

The junior officer and the men who moved the gold and sealed the cave would of course have to be silenced permanently. And the inquisitive corporal, as well.

During the first few years in Argentina he thought about the cave frequently, especially when he read newspaper stories announcing that more "Nazi plunder" had been discovered in Austrian lakes and in the salt mine near Salzburg. What stupid fools those other generals had been! One of the accounts had read, "Additional loot is said to have been ditched in a Tirolean lake near a castle, but a search of the lake bottom was unsuccessful." Apparently they had given up.

He had been astonished at how soon Europe recovered and began to prosper. Eventually he wished he hadn't done such a thorough job. If only he hadn't buried the entrance under a ton of stone, he might have been able to go back some day. . . .

It was a wildly tempting thought. Krueger did not need the money. Margarita's father had been a wealthy Argentine businessman. But he had felt since the war that his life was unfulfilled. What good he might have done with so much gold: campaigns and candidates to back, movements to finance—and he could have bought an influential newspaper, put the right kind of courageous editor in charge. With large amounts of money it is possible to leave one's imprint on the world. Men like Hitler had started with far less.

But the gold was buried under tons of rocks. The cave could only be reopened with heavy machinery, there would be dozens of gawking villagers standing by waiting to see the results. Impossible! He wiped the vision out of his mind.

The years hurried by. He forgot about the cave for long periods of time, until the early 1970s, when he watched the price of gold spiral up with an exquisite mixture of pain and chagrin. At times he felt as though the Western monetary crisis had been concocted solely as a device to torture him. He estimated the weight of the gold in the boxes and calculated its value at the old rate of $32 an ounce as against the new rates of $100 and then $130 and $145 an ounce. Helplessly he watched the treasure rise in value from several hundred thousand dollars to well over a million.

Then one day in 1973 he was leafing through a travel magazine in his dentist's waiting room. He idly flicked past a picture of a grinning couple in bathing suits who were sitting on a rock ledge with their feet dangling in water. Their faces and the rock walls around them were tinted a pale blue, just like the walls of the cave. He flipped back to the picture quickly, with a kind of foreboding.

He already knew that it wasn't the same pool, of course. This one was in the Caribbean. The text said that the blue reflections were caused by sunlight bouncing in from the ocean outside through a water-filled tunnel running into the cave's pool.

The significance of the matter-of-fact explanation swept over him. It meant there must have been another entrance to the cave in Tellin, through the lake into the pool. He felt like an idiot. At that very moment someone might be cramming his pockets full of *his* gold and swimming out with it to the surface of the lake.

He shot to his feet, feeling ill, and left without saying anything to the startled receptionist.

After he had calmed down he recognized that it was very unlikely that anyone had discovered that entrance to the tunnel, the second entrance that must lead from the sunlit Tellin lake into the cave. If the Americans hadn't found it when they searched the lake, it wasn't likely anyone else would stumble onto it. It was probably very well hidden. He conjured up that scene outside the obliterated cave door. The lake was no more than twenty meters away from where he stood. Not too far to swim underwater. But even if the outer entrance were near the surface of the lake, it would be suicidal for anyone, even an expert swimmer, to enter an unexplored tunnel without diving equipment. It wasn't likely that anyone would try, especially since no one knew where the gold actually was.

After he reassured himself, there was something else to think about. This belated discovery of the properties of refracted light revived his dreams of getting the gold for himself. He soon realized that this too was an unrealistic idea.

It would take time to locate the tunnel, more time to bring the gold out bit by bit. Everyone in the village would remember how the American soldiers had searched for the missing treasure. A stranger coming to that out-of-the-way lake with diving equipment would arouse suspicion. For all he knew the authorities might still be watching Tellin, waiting for someone to come, even at this late date. They might even be waiting specifically for *him*. It was too risky. The discovery of a second entrance to the cave changed nothing. The gold was as inaccessible as ever.

Until today. Until he had learned that some insignificant fisherman had caught his nets on wooden pilings in the lake. He stared at that last sentence. They would be going to the lake that summer . . . "if funding can be obtained and if a qualified crew of young volunteer divers can be assembled."

Around ten o'clock, as he walked back toward his cottage along the sidewalk curving around the bay, Ortega passed the Acapulco Ski Club. He had passed the plain white building countless times before, but this time he took a real interest in the billboard announcing daily lessons in scuba diving.

Yes, he decided, it just might work. He was too old now to do it himself of course, even if he were not afraid of being recognized. However, it was fortunate that he and Margarita had a child. A child now grown, who was strong-willed and obedient.

And it occurred to him that there would be another ally waiting in Tellin, though a reluctant one. Someone who, because of a shameful secret, could be pressured into helping in every way necessary to make the plan succeed.

He felt buoyant and excited, as he had not felt in years, not since those mornings when he was about to send his troops into battle. Striding briskly for a man his age, he passed the street he ordinarily took to go to the native market. He had completely forgotten about buying provisions for his evening meal. His mind had marched ahead to the coming summer; there were only a few months in which to lay the groundwork.

But first there was an overseas telephone call to make.

ONE

Like most twenty-two-year-old Americans, Judy Hamilton didn't like phonies. She liked stool pigeons even less, and that's why she hated doing what she had decided to do.

She was sitting in a decompression chamber ten feet below the surface of the water. It was a yellow metal ball filled with air and large enough to hold four people sitting knee to knee. At the moment she was alone.

The others on the afternoon dive had just gone up. They had finished their decompression time ahead of her. She would have to wait in the chamber for another fifteen minutes to avoid any danger of getting "bent." There was a paperback book in the hollow behind the semicircular plastic bench but she didn't feel like reading. The final straw that morning had irritated and depressed her.

The cardinal rule of scuba diving is "Never dive alone." Divers are supposed to use the buddy system, always swimming in pairs. Yet someone in the Tellin expedition had been breaking that rule off and on all summer.

The first time it happened Judy was busy doing her job, setting up one of the grids over the dwelling site on the lake bottom, when she looked up and found her "buddy" was nowhere to be seen. When she complained about it topside, the answer was a friendly apology and a promise that it wouldn't happen again.

They hadn't been paired off again for eight days. That was the way the rotation of assignments went. Working together in the soft lake bed, they had pulled out a flat rock blackened on one side—Judy thought it might have been a hearthstone—and sent it up in the lift. When they swam back to the site, Judy got involved in uncovering another stone. Within a couple of minutes she realized she was alone again. The water was moderately cloudy; it was impossible to see more than a few yards in any direction. Still she thought she caught a glimpse of a wet-suited figure moving off toward the castle shore just as she glanced that way.

That time the excuse was a not too polite version of "mind your own business."

But it *is* my business, she had protested. If something should go wrong, I might need to share your hose to get to the top. Besides, why should I do all the work while you're off joy riding over there around the castle? Why don't you do that on your own time? But of course she knew the answer to that last one: Professor Windle wouldn't have allowed any solo diving.

And so there was another apology, a cooler one this time. *Sorry, but I was sure I saw some more pilings over that way. I thought it might be a new section of the lake dwelling.* Judy didn't believe a word of it.

A couple of weeks went by with no problems, maybe because they had all been diving together for a change. But then that morning the same thing happened again, and so at lunch she gave due warning. "As soon as Professor Windle gets back from the village I'm going to tell him all about this."

She realized now she hadn't been happy since she came to Tellin. It should have been fun. The Bronze Age period was interesting to her. Besides, there were four other divers about her age on this dig, all but one of them easy to get along with, and Professor Windle was okay. Despite that absent-minded way of his, she knew he was a good archaeologist. And they got to bunk in the old castle, the only place in the village large enough to accommodate all of them and give them a work area for the cataloguing. The Alpine countryside was picturesque too, what she had seen of it.

And yet there was something vaguely off-key about the whole expedition. Too much money was involved, for one thing. On the two other underwater digs she'd been on, the student divers had had to *pay* for the privilege of working their tails off all summer. And that was the way it usually was, because the foundations that financed these things didn't exactly have money coming out of their ears these days.

But at Tellin the divers got, of all things, a small salary. What's more, the expensive equipment on the boat was brand new. All very nice of course, but somehow it didn't seem quite right. Despite all the hard work, it seemed more of a lark than a serious archaeological expedition.

The chamber she was sitting in was one example of the expedition's affluence. Ordinarily divers had to hang on ropes below the diving boat to wait out their decompression time. In comparison, the spartan metal chamber was a Rolls-Royce. Since it was kept filled with air from a hose on the boat, the people inside could dry off and sit out their waiting period in relative comfort.

The ball-shaped chamber was held at ten feet below the surface by a steel cable that ran from the bottom, down and through a hook embedded in a concrete slab on the lake bottom, and up again to a winch on the boat. If it weren't secured in this way, the air inside, which was compressed naturally by the surrounding water, would make the chamber bob explosively to the surface, and that would be dangerous to anyone inside.

The decompression chamber was also a reminder that diving to depths of sixty feet or more is not a natural activity for human beings. At that depth the air a diver breathes from his tank is at a pressure of around forty-five pounds per square inch or about three times the normal atmosphere. If the air weren't pressurized, the lungs would collapse from the unequal force of the surrounding water.

The body could adapt to that pressure; the real problem came when the diver went back to the surface. If he didn't take it slow to give himself time to breathe off that compressed air in his system, nitrogen would form dangerous bubbles in the blood, just like the bubbles in a champagne bottle after the cork is popped. The result could be decompression sickness, the bends, and either temporary or permanent paralysis, even death.

But getting bent wasn't the worst thing that could happen. If the diver came up extremely fast and didn't remember to breathe out on the way up, the compressed air inside the lungs would expand rapidly as it met less and less counteracting pressure from the water around him. The deeper the swimmer was and the faster he went up, the greater the danger would be. The lungs could be blown up until they burst like balloons.

She often thought, when she was going down to the bottom of the lake, that it was like entering an upside-down pyramid which got narrower and narrower as she descended. And then, coming up again to the broad flat surface of the lake, everything widened out, expanded, as the water grew lighter and lighter, not only in terms of vision but also in terms of weight.

That's why it was easy to remember the twin dangers of diving—the nitrogen expanding in the bloodstream, the air expanding in the lungs—as she rose to the surface that was the wide upside-down base of the pyramid.

Judy knew, however, that both these dangers could easily be prevented with a few sensible precautions. One simply had to be patient and take the proper amount of time getting to the top. Diving could be almost as safe as crossing the street if one merely used one's head. That's why it was especially irritating to her when someone deliberately broke the rules.

The water in the open hatch at her feet was gushing a little. She'd never noticed anything like that before. Possibly a large boat had passed overhead, though she had never noticed any big boats out on the lake. Anyway.

She ran her hands through her short damp hair and leaned back against the cool metal wall. She couldn't make up her mind whether to leave the expedition or not. Professor Windle wouldn't like it if she left before the project was finished, but what she was going to tell him was bound to cause trouble. It was hard enough for a group of people to live and work together twenty-four hours a day without the added complication of a dispute like this.

She wished Ann were there to talk to. Her sister had helped her solve much bigger problems than this one simply by listening and asking the right questions. They would talk it over and somehow the right thing to do would become clear. They would usually end up laughing about whatever it was.

When they were younger Judy had been the rebellious one who came in late or broke the other rules their loving but rather strict father had laid down. Ann had acted as a mediator between them, soothing both their hurt feelings.

In high school Judy had been hung up on astrology for a while and it seemed to her that she and Ann were good examples that there was something to it. Judy was an Aries, "headstrong and impulsive." Ann, on the other hand, was a Libra—"a natural peacemaker," the book said, and "the most agreeable and pleasant sort of friend."

Except that Ann wasn't as placid as Libras were supposed to be. There was something of an Aquarius about her—intelligent and generous, with a good sense of humor. And every once in a while she could be as tough as the Capricorn mountain goat when she was certain she was right about something. She had a temper too, though she didn't use it much. Naturally Ann didn't take astrology seriously. "People are much more unpredictable than that," she said.

It was ironic, Judy realized, that of the two of them it was Ann who had chosen the wrong man to marry. She ordinarily was a good judge of people, yet she had fallen in love with Don Cole. Judy herself had thought when she met him that he was a nice guy. He was certainly charming. Then she saw how, shortly after the marriage, her sister seemed unhappy and preoccupied—a mood not at all like her. They must have been having their problems even then, although Ann never talked about it.

Then during the last year she found out from someone who worked with Don in New York that Don had been having occasional affairs with other women. Ann must have known.

It wasn't fair. Ann was pretty and bright and giving. She deserved the best.

Judy rezipped her wet suit. It was almost time to go up. She decided she would go to the village as soon as she got dressed and try to call her sister.

It must be about nine o'clock in New York; Ann would have already left for work. She would call her at the office and ask her to fly over for a couple of weeks if she could take some time off, and then they could go back to the States together. It would be good for Ann to get away for a while. She had sounded very low the last time Judy had talked to her. That was a couple of weeks ago, just before the divorce went through.

She noticed that the water was still churning a little. A few drops lapped up over the rim of the hatch. She found herself thinking, *What are those guys doing up there?* She got the weird sensation that the chamber was dropping very slowly to a lower depth. *If they don't stop fooling around, I'll have to do my decomp time all over again.* But that was nonsense. They never lowered the chamber, and there was really no way to detect movement in that metal cocoon. It was undoubtedly an illusion of some kind.

But the water fell away with suddenness. With steadily mounting momentum the steel chamber drove upward toward the light.

Professor Windle had been in the boat's cabin writing a letter to his wife. One of the major reasons he had been enthusiastic about going to Tellin was the opportunity it presented to get away from the gentle nagging of his wife Lucille, but now that he was away he missed her terribly. *Our project is going well,* he wrote, *and we should be able to finish up by . . .* He stopped, frowned, and after considering for a moment crossed out the last part. There was no point in indicating when he might be coming home. If things should change, he would have to go into elaborate explanations to keep her from asking, "But I thought you said—?" Now he would have to copy the page over.

He became aware of a constant mechanical shrieking outside the cabin, then someone running, then a grating clatter. *What's going on?*

He spun around to look out the window just as the yellow ball exploded out of the water. The pen in his hand fell to the floor, he ran outside to grasp the starboard rail and stare. Before he could think to bark an order he saw that one of the divers was in the water pulling the yellow chamber alongside.

The young girl was half in, half out of the hatch as though she had realized too late what was happening to her and tried to get out. It was not a good sign. If she panicked and forgot to breathe out . . . The other three divers crowded along the rail and helped the boy in the water lift her up and lay her on the deck. Eyes open, her face had frozen into a photograph of her alarm.

It was no use trying to administer life-giving breath into the broken lungs.

TWO

Ann Hamilton Cole spread the road map on the hood of the Volkswagen and tried to locate her position. On paper Austria was a plump tadpole struggling to escape eastward from under the boot heel of Germany. Vienna was its large round eye and the Alpine province of Tirol, where she now stood, its tail. With a forefinger she traced the emphatic red line of the main highway west from Innsbruck through Zirl and ultimately onto the thinner blues and blacks of the secondary and tertiary roads she had been following since two o'clock. She marked what she guessed to be her approximate location with a fingernail and calculated that she was still about twenty miles from Tellin.

She breathed out heavily and looked around her. During the long drive she had gotten used to the scenery. The road zigzagged through a deeply cut valley. The horizon was blocked on either side by jagged gray mountains, and the valley steadily rose toward another line of mountains in the distance. On the nearer slopes dark feather-shaped evergreens marched down to a patchwork of rolling meadow. Up ahead there was a whitewashed farmhouse with two golden horses grazing near its back door.

She wasn't impressed. In her mood, the scene looked too much like a cheap print, the kind they sell in Woolworth's. "The Tirol in August," they would call it.

She brushed a strand of hair off her forehead and squinted in the direction of the farmhouse, thinking, *What a place to have car trouble.* She weighed the pros and cons of walking to the house for help. There was probably no telephone, and no nearby garage to call in any case. Maybe her German wasn't good enough to make herself understood.

She was refolding the map when the distant hum of an engine made her look up. An automobile, hardly more than a spot on the ribbon of road, was moving rapidly toward her. With relief she stepped onto the edge of the unpaved road, determined to flag the car down if she had to.

She waited impatiently while the car alternately rose and sank behind dips in the road like a ship riding out a storm. As it came closer she saw it was the black Mercedes she had seen earlier that day when she had stopped at a little roadside café near Zirl, and her relief became mixed with a twinge of apprehension.

She had stopped to have a cup of coffee and to ask directions. There hadn't been any other customers. She sat alone at one of the tiny marble-topped tables and thought about Judy.

When she walked out the front door a few minutes later she saw the black Mercedes. What struck her at once was that in all the vacant parking area it had been put in the slot directly next to hers. And the driver was nowhere in sight. Odd, she had thought. And then she had spotted him.

He stood up from just behind the rear bumper of her rented car. His right hand went to the pocket of his windbreaker jacket as though he had been stooping down to retrieve something he had dropped. One step put him between the Volkswagen and his own car. The motion looked perfectly natural, and yet she was vaguely annoyed.

The driver of the other car certainly hadn't looked like a suspicious character. Tall and neatly dressed, he had the same healthy tan and sun-bleached blond hair that the mountain guides had in those Austrian travel brochures she had picked up at Kennedy Airport. His face was the type everyone automatically pegged as "honest"—a strong square jaw, high cheekbones, straight nose, eyes set wide apart under a moderately high forehead.

He patted his pocket again, apparently without thinking and started toward her, striding with an easy self-confidence. She put his age in the middle thirties. As their paths crossed his eyes flicked across her face without curiosity, and he went on into the café behind her.

It wasn't until she had unlocked the VW and started the engine that it came to her exactly what had been wrong. The cars were side by side on the parking strip and parallel to the road, with the Mercedes nearer the building. In order to get to the spot where she had first seen him, he would have had to get out and walk *away* from the building, all the way around the rear of his car.

Probably, she told herself, he had gone back to put something into the trunk or to check his tires, and had dropped his keys. Still, she didn't like it.

Momentarily irritated, she had taken a minute to check her belongings. The yellow overnight case she had borrowed from her roommate was still on the back seat. (During the divorce she hadn't asked for anything, not even the luggage that was a wedding present. Instead she got a job with a textbook publisher and found someone to share the rent in a new apartment.)

The only thing that couldn't be replaced she carried in her purse. It was the letter Judy had written the day before she died. There was enough in it to make her wonder if her sister's death had really been an unavoidable accident.

A dozen yards down the road the black Mercedes whined into a lower gear and decelerated. The driver turned his head her way and nodded deliberately before wheeling off into the grass and cutting his engine. A moment later he was walking back toward her with a smile and a courteous but appraising look. She could almost hear him sizing her up in his mind: age about twenty-five, slender, five-foot seven, short auburn hair, heartshaped face, and—getting closer now—dark green eyes.

Had she been more self-centered she might have realized that he was also thinking, "Quite good-looking." She reminded him of a French actress whose name he couldn't remember. Stopping in front of her, he took off his dark glasses so they could face one another directly. "I'm about to ask you an obvious question," he said, in unaccented English.

But she didn't wait for him to ask it. "Yes, car trouble. The motor made a funny sound a while back. A warning light came on on the instrument panel. When I stopped the engine seemed very hot, so I was afraid to try to start it again." She wished she weren't so ignorant about automobiles.

She followed him to the rear of the little car and watched as he bent down to look under the hood. He sighed, then reached into the bottom of the compartment and lifted out a length of flexible black rubber. "Fan belt's broken. Do you have a spare?"

She shook her head. He stood up and looked up the road. He said, "There is a part-time mechanic in Tellin that may have one. Unfortunately, this happens to be a feast day and he may not be at home." He turned back to her. He was wearing a gray turtleneck under his windbreaker, and she noticed when he turned around that his eyes were the exact color of the sky behind him. "Why don't you lock your car and let me take you wherever you're headed. You could arrange to have the man come out and install the belt tomorrow."

It was the logical thing to do, yet she had misgivings. She didn't like the coincidence of running into this stranger again, no matter how pleasant he seemed. Stalling for time, she glanced back the way she had come and saw an old man in an ox-drawn cart making his way slowly toward them. The cart was loaded with hay. She noticed that the sun had already slipped behind the mountains. Shadows had swept across the road and were lapping at the edge of the farmyard. When she faced him again he was watching her patiently, a little amused.

"Or perhaps," he said softly, "I could ask the farmer there to give you a ride in his cart."

She broke into a smile in spite of herself. "That won't be necessary, thanks." She went to get her luggage.

They didn't introduce themselves until after they had gotten settled in the big car and started off. He was Alex Schuler, he said.

"Ann Cole. From New York." She had decided not to go back to her maiden name despite the fact that they hadn't had any children. To drop Don's name, she felt, would be like trying to pretend the marriage never happened. "Do you live in Tellin?"

"Just for the time being. I come here every summer, lately."

"Then you're on vacation?" She watched his face while he drove. It was an intelligent face, not blandly handsome like the one in the travel folder.

"I wish I were. I'm finishing up a research project. I'm a limnologist."

"Limnologist," she repeated. I ought to know that word, she thought. It was part of her job to look up unfamiliar words and make sure the writers for the company she worked for hadn't misused them. Sometimes their authors, even the university people, would make some real bloopers. It was up to assistant editors like her to catch them before they got into print. She thought that "limn" meant to draw or paint, but Alex Schuler didn't look offbeat enough to be an artist.

"It's a branch of biology," he said. "The study of freshwater lakes and ponds. The project idea was to take the most secluded lake I could find and try to discover how much damage pollution has done to it."

So he was working in the lake where Judy died. She managed to say, "That must be interesting work."

"I suppose. It's probably duller than it sounds. Collecting water samples from various parts of the lake, throwing chemicals in to see what the reaction is. Keeping endless records."

His being a scientist fit, she decided. He looked coolheaded and reasonable. She could imagine him very much at home in a lab, fiddling with test tubes and making his coffee on a Bunsen burner. After a small silence she said, "By the way, how did you know I was an American?"

He looked at her, giving an ironic smile. "How did you know I knew you were an American?"

She smiled back. "I assumed it, the way you spoke English with me right away."

"Of course. When I saw you earlier at that coffee house near Zirl, you looked very American to me."

The way he casually mentioned their earlier encounter startled her at first, but then made her feel more relaxed. Apparently he had nothing to hide. "I'm not sure how to take that," she said.

"I meant it as a compliment. Americans aren't afraid to look at strangers directly, I've found. We Europeans are usually more guarded, more closed in on ourselves." He was like that in a way, she thought. Very European. He was being friendly to put her at ease, but she sensed that he was holding back somehow, more than an American might have. Not that she minded. It suited her mood.

He said, "But of course it's foolish to think in stereotypes. I jumped to the conclusion you were a one-language American. Not that you look like a typical tourist," he continued, giving her a sidelong glance. She had on a charcoal suit that wasn't as expensive as it looked and a white blouse, not the kind of thing she would have worn if she had come to Austria on vacation. "So you do speak German, after all?"

"Not very well, actually. My grandmother was German, and we used to hear it spoken around the house."

"We?"

"My sister and I," she said quietly. "I suppose I jumped to conclusions myself. I didn't expect you to speak English so well. You don't even have an accent."

"I spent a year in Connecticut as an exchange student years ago. I stayed with a very nice family, a doctor and his wife with two kids. I had to learn the American language fast to keep up with them. But I have a feeling I'm still using slang that went out of style about ten years ago."

"Did you like America?"

"I loved it. We went camping once in the Adirondacks, I remember, when the leaves were just starting to turn. Beautiful country. The main thing that impressed me, though, was the friendliness of the people, and their optimism. I liked it so well that when my time was up I got a job and stayed on an extra six months. Before I left I rode a Greyhound bus out to San Francisco, saw the Grand Canyon."

"Have you been back since then?"

"Only once, a couple of years ago. For a scientific meeting in Washington."

"You gave a paper?"

"Yes." She could imagine that too, his standing at a lectern in a large auditorium while he explained the technical slides that flashed on a screen behind him. He would probably look very authoritative in a suit and tie with that square jaw of his.

They had gone several miles and hadn't passed a single car. She said, "It's lucky for me you happened to come along when you did." She deliberately made it sound like a question, and he picked it up. She was curious about why he'd been driving on the road from Zirl.

"I had to go to Vienna this weekend," he said.

It was over 250 miles to Vienna. Must have a special girl friend there, she decided. Or a wife. "Did you have a good time?"

"It was pleasant enough."

"It rather startled me, you know, when I first saw you in that parking lot. All of a sudden there you were." She tried to sound extremely unconcerned.

He reached up to adjust his sunglasses without taking his eyes off the road. He looked unhappy, she thought. "I'm sorry if I gave you a scare. I was afraid I might have. I thought I recognized your car. A friend of mine owns a blue Volkswagen. It wasn't until I checked the license plate that I saw I'd made a mistake. And that's when you came outside."

It was plausible. There were a lot of blue Volkswagens in Austria. As though to change the subject, he said, "May I ask why you're on your way to Tellin? We don't see many tourists this far off the main roads."

Now it was her turn to explain. What could she say? That she had come to find out why and how her sister had died? She didn't want to tell him that. "There's an archaeological expedition at the lake," she began. "They're staying at the castle in Tellin. You must know about it?"

"Yes." He was clearly waiting for her to go on.

"I'm interested in finding out more about it." She hated lies. She had heard nothing but lies for the past year, or so it seemed to her. That had been the worst part of it.

"Are you an archaeologist, Ann?"

"No, I work for a book publisher." It would have been easy to go on and maintain that she was planning to write a book about the expedition or something, but she didn't.

"Oh, I *see*," he said with a faint smile. What he meant was that her explanation hadn't cleared up anything but he could tell that she didn't want to talk about it.

They rounded a gentle curve and he said, "There it is now up ahead. *Schloss* Adler. Eagle's Castle."

It had slipped into view from behind a hill they had just passed. Crowning a steep-sided offshoot of the mountain behind it, the castle was an imposing fortress with pewter-colored stone walls and a single round tower. Ann's first thought was that it didn't look like anything out of a fairy tale. In fact, it reminded her of a prison.

"Looks intimidating, doesn't it?" Alex remarked. "It must have been built to look that way to discourage the peasantry from thinking about attacking it. It dates back to the sixteenth century, I believe."

She already knew that. When Judy wrote her about living in the castle she had made a point of looking it up the next time she went to the New York Public Library. She had found an old leather-bound book that told about Schloss Adler. It had been started in 1580 and completed in 1592, twelve years being a not unreasonable length of time to complete a structure like this one. During those years when the peasant laborers were struggling with the stone building blocks, Sir Francis Drake was exploring the New World, the Spanish Armada was destroyed, and a novice playwright from Stratford-on-Avon arrived in London.

"It's owned by a baron, isn't it?" Judy had mentioned him, saying that he was "weird."

"Wilfred von Toblen. One of his ancestors supposedly got the castle by helping the Hapsburgs put down a rebellion."

"Do you know him?" She took her eyes off the castle for a moment to study Alex's face. She saw now that there was a tenseness around his eyes and mouth that she hadn't noticed before.

"No," he said. "He's kept pretty much to himself since the war. After the Nazis took over, he and his wife were put in prison for about a year as 'political unreliables.' The villagers say he hasn't been the same since. They respect him very highly, though, because he was an anti-Nazi even before the war. The Germans had sympathizers in many parts of Austria then, you know, but not in Tellin. Their soldiers were stationed here for a while and the people got to know them too well."

Then he asked, "Where are you staying?"

"I haven't made arrangements, but I understand they take in guests at the castle. I was hoping they'd have room for me there."

There was a distinct pause. "You must have decided to make this trip rather suddenly, I take it."

She realized that it must sound odd to him, her not making reservations anywhere. "Yes, it was sudden."

She could see the village now—a small neat group of houses and the long thin steeple of a church. The buildings were huddled together as though they had slid down the mountainside and ended up clustered along the road. As they reached the outskirts, Alex geared down, and the sound of the motor changed its pitch.

The houses were all two or three stories high.

The bottom walls were whitewashed stone, the upper floors mostly dark unpainted wood under broad overhanging eaves. She saw wooden balconies covered with ivy and lined with boxes of red geraniums. Off the main street brief cobbled alleys twisted up toward the mountains on her right and down to the lake on the other side. Under other circumstances she might have thought of the village as idyllic. It was the type of place the travel agents called "unspoiled."

The road widened to circle around a fountain with a statue of a robed saint she didn't recognize. If space hadn't been at a premium, there might have been a full-fledged square here in the center of town. As they passed by she noticed a little *Gasthaus*. A thin man wearing a long apron and baggy trousers was clearing an outdoor table under a striped awning. Farther down the street a pigtailed girl in a print dress was sweeping a front stoop. Ahead of them the castle and its hill dominated the village and created the illusion that it formed a high wall at the end of the street. It was getting dark.

"The castle is farther away than it looks," Alex told her, "and it's possible they may not have a vacancy. Are you sure you want to go up there this evening?"

As a matter of fact, she didn't. She was dead tired, and she had gotten the front of her suit dirty leaning against the side of her car to look at the map. Better to wait until morning to get a fresh start, if she could. So she asked, "Is there an inn in town where I could stay overnight?"

"There is, but I know an even better place," he said cheerfully. "A very nice woman named Elsa Bruner rents out one of her upstairs bedrooms. She lives in that little house just ahead. Her husband is a friend of mine. She'll be glad to put you up."

It was almost as though he had been expecting her.

THREE

In her sixties, Frau Bruner was a plump red-cheeked woman with the hard rough hands of a farmer's wife. Her upstairs room was small but comfortable. Ann liked it on sight—it didn't pretend to be anything more than what it was, a family bedroom occasionally let out to visitors.

The wide pine floorboards had been scrubbed until they were white and smooth. There was a small crucifix and a photo grouping of the Bruner family on the wall. Ann's glance moved past the photographs to the tiny balcony facing the lake. She said that the room would be fine, that she'd probably be staying just the one night.

Alex had brought her suitcase up. After Frau Bruner went out he said, "The *Gasthaus* will be serving dinner in about an hour and a half. I'd like it very much if you'd join me."

"Thanks, Alex, but it's been a long trip for me.

The flight from New York and then the drive here." She let her voice trail off.

He smiled. "I won't keep you out late, I promise. Just a quiet dinner."

She was tempted. It didn't seem right to say no after he'd been so nice. "All right."

"Good. I'll see if I can find the mechanic before I pick you up."

"Thank you, Alex."

"It's no bother."

His smile was really quite disarming, she thought as she watched him leave. But that tenseness was still there around the eyes.

The first thing she did after he left was take a long hot bath in the old-fashioned bathroom down the hall. While she was getting dressed she studied the pictures on the wall. A younger and slimmer version of Frau Bruner beamed at her from across the years. She was in her wedding dress, sitting in a buggy trimmed in white that was parked in front of the village church. The groom at her side looked sternly self-conscious in a dark suit. There was also a series of photos of a young girl in small round frames.

The girl resembled her father. Her life was recorded from babyhood to about sixteen, then stopped. Even the most recent picture seemed to be several decades old. She wondered if the daughter had gone away for some reason, or died.

She was very aware that it was possible for a girl, even a girl loved by her parents, to die young. She thought of Judy again.

She put on a blue skirt and blouse and brushed

her hair. She had cut her hair very short so she wouldn't have to spend much time on it, and she was lucky, she simply had to run a brush through it and it fell into soft curls. Now she brushed it back off her face and let it go at that.

Because she had a few minutes, she took the letters out of her purse to read them again. They were the last two letters Judy wrote her, and she had read them so many times recently that the paper was flimsy along the creases. The first one was dated six weeks ago. She never read it without feeling wrenched by how alive Judy had sounded. The handwriting was angular and slanted sharply to the right.

> . . . We're finally going to get started on the serious work tomorrow. We're all pretty well settled at the castle now. (That really sounds super, doesn't it? Living in a castle, I mean. But sooner or later you realize that despite the name it's just a drafty old building.) Our rooms look like they were intended for medieval monks they're so small. About the only halfway romantic part is the tower, and we're using that as a work area.
>
> An old baron owns the place and lives here with his wife. They must need the money they get from renting rooms. He's a real character, kind of weird. . . .
>
> My roommate is Karen Phillips, who's from Florida. Oddly enough, all the other divers are from Latin America. That's one funny thing about this dig. Professor Windle said

the foundation that's sponsoring us is backed by someone who wants to provide training for archaeology students from Spanish-speaking countries. It's called the Omega Foundation—have you ever heard of it? Let me know if you read anything about it. It's so new I don't know much about it myself, and I'm curious.

Mother wrote me about you and Don. I know you must be sad about the divorce, Ann, but I honestly think it's for the best. . . .

For the best. Everyone was very sympathetic. One of Don's friends told her, *He knows he's acted like a heel. It isn't that he doesn't love you.*

How had things gone wrong? She thought she knew him well before they married, she thought they shared the same values. She met him in college when he was a law student, and after the wedding they had moved almost immediately to New York. It surprised her to learn how important it was to him to get to the top at his job with the law firm there, not just to make a name for himself but for the money it would mean. He had always insisted he didn't care about money. Now he talked of "attracting the right clients," meaning the wealthy and influential ones. That was when it started to go wrong between them, not later. *She didn't mean anything to me, don't you understand? It happens all the time.*

Okay, so maybe I'm living in the wrong century.

"Stupid." She said it out loud and got abruptly

to her feet. It was no use going over it again and again. It was finished.

She looked at the date on the letter. Judy had called her from Tellin two or three weeks after that. They had both made an effort to sound cheerful but without much success. Judy had seemed vaguely troubled, but there was nothing specific she could remember now. Ann thought at the time her sister might have been worried about how she would react to the divorce.

The other call from Austria came two weeks later, but it wasn't from Judy. The woman at the other end read the message impassively. "We sincerely regret that we must inform you of the death of your sister, Judith Hamilton. . . ." There was something about an accident at the archaeological site.

It took almost an hour to reach the man who had sent the message, a police official in a large town near Tellin. He tried to explain how it happened. He told her the winch had failed, allowing the chamber to shoot up to the surface unexpectedly and causing a rupture of the lungs. He patiently answered her questions. She told him she still didn't understand. He told her he had written a letter giving all of the details. "Diving is a dangerous occupation," he said. "I'm very sorry."

When the last letter from Judy arrived two days later she had a momentary sunburst of hope that there had been some horrible mistake and Judy was alive after all. But of course the postmark ended that. The letter had been mailed the day before the accident.

She went out onto the little balcony. The lake was dark and calm. Since the Bruner house was at the far edge of the village, the castle loomed over it like a black mountain. It was chilly outside.

The policeman's letter, when it came, hadn't really told her much more than what he said to her over the phone. There were so many unanswered questions, but she wondered now if she had done the right thing to come here. She went back into the bedroom, picked up Judy's final letter, and read it through for what must have been the twentieth time:

Dear Ann,

We've been pulling a good number of artifacts out of the lake, and the weather here has been beautiful. I suppose I should be feeling good, but somehow I'm not. Actually, I'm thinking about coming home for the rest of the summer.

I suppose that deserves some kind of explanation, doesn't it, since I was so enthusiastic about this expedition before I left? The thing is, the other four divers I'm working with are all pretty inexperienced, and it's getting on my nerves. One in particular insists on wandering off when we're supposed to be diving as a team, and that could be strictly bad news if anything ever happened to my tank. But you mustn't worry about me, because I've already decided I'll have to tell Professor Windle about it if it happens again, and he'll take care of it. I hate to do that, though, because it might get this person canned. You know me, I hate hassles, so don't be surprised if I quit. I'd really like to see you, anyway.

Let me know how you are getting along, Ann. I have to run now to get this mailed.

Love, Judy

It had happened the day after that. The next afternoon she was killed.

She put the letters back in her purse when she heard Frau Bruner call her name from the bottom of the stairs. Alex was waiting.

FOUR

The *Gasthaus* wasn't what she had expected. It had the homey atmosphere of a neighborhood café. There were no touristy decorations, only a bare wooden floor, tables with gingham cloths, and a long counter behind which a man with white hair and a mustache dispensed food and drink. Three men dressed in rumpled black suits and white shirts were at one of the front tables. In one corner there was a glazed earthenware stove set on a stone foundation. Alex led her to a table by a window in the back, where they could look out on a small garden.

"We're in luck, by the way," he said as they

were getting settled. "The mechanic will have your car ready by tomorrow morning." He had put on a plain gray sports coat over his turtle-neck.

"That's good. Did you have any trouble finding him?"

"No, he was at home after all. He lives near the Bruners in that white house with the big barn. He was very happy to get the work."

The waiter brought the menus. Alex left his lying on the table while she studied hers. "Avoid the schnitzel," he advised her. "Everything else is good."

She broke into a smile. "That reminded me of something. Not long ago at my job I read a manuscript by a man who's supposed to be an authority on English usage. You know what he wrote? 'Avoid clichés like the plague.' "

He laughed. "I'm glad I didn't say that. I might have, because I'm not always aware of clichés in English. He couldn't have been joking?"

"Not likely. The rest of the book was very straight-faced and dry." She decided on the *Gröstl*, beef roasted with thinly sliced potatoes, onions, and spices.

The waiter came back and took their order. The menu gone, she folded her hands on the tablecloth. Her first uneasy impression of Alex at the roadside café had been wiped out now by a much more favorable one. She found him intelligent and warm, very likable.

A moment later he said, "You know, I've been trying to figure something out."

"What's that?" she asked.

He leaned over and touched the ring finger of her left hand. After she and Don separated she spent a lot of time lying in the sun trying not to think. She hadn't taken the wedding ring off until the day the final decree arrived in the mail. Now the white strip of skin was like a ghost of a wedding band she had to carry around with her.

"I'm divorced," she said. She realized that it was the first time she had said it.

"Very recently, it must have been."

"A few weeks."

"Sorry."

"Don't be. I'm not." She caught herself and added, "No, that's not true. I am sorry, but it couldn't be helped."

"I think I understand. My parents were divorced, years ago. Luckily it was a friendly breakup. No bitterness. Having two temperamental people in one household just didn't work out very well. My mother happens to be a soprano, you see, and my father was a conductor. They worked for the same music company in Vienna. But they remained good friends, right up until my father died a few years ago."

It was a background that surprised her, and yet somehow it seemed to fit. "It isn't like that in my case, I'm afraid." During the last few months she and Don had quarreled so often and so bitterly that she sometimes felt as if she were clinging by her fingernails to a sheer wall of glass.

He must have read some of that in her face. "Don't worry. It will fade," he said gently, and she knew he was referring both to the suntan and the bad memories.

"And what about you? Are you married?"

"No, I never have been."

She wondered why. Her first and continuing impression of him—based mostly on the forthright way he looked at her—was that he definitely enjoyed women. She decided that perhaps Alex was overly cautious because of his parents' experience.

"But you must have come close to it," she observed matter-of-factly.

"Yes, close to it. A couple of times." Whatever had happened, it evidently wasn't anything that haunted him.

"Did you inherit any of your parents' talent for music?"

"Their genes seem to have canceled each other out. I'm afraid I'm strictly a listener. They tried to give me violin lessons when I was growing up, but the teacher kept picking up the metronome as though he wanted to throw it at me."

She smiled at that. One thing she couldn't picture was Alex playing the violin. "Somehow I can't imagine you being temperamental either, Alex, as you say your parents were. The genes must have canceled out there too."

"Maybe," he said mildly. "Or maybe you simply don't know me well enough yet."

That "yet" was like a promise of possibilities between them. She glanced down at the tablecloth. That wasn't what she had in mind at all. There was an abrupt pause in their conversation.

The food came and they ate. The *Gröstl* was not at all bad. He ordered coffee, black for him and *mit Schlag*—a glass mug with a great mound of whipped cream on top—for her.

Over coffee she asked, "Do you know any of the people staying at the castle?"

"I think I've met them all at one time or another. When I'm out on the lake I often pass their boat. I've even been up there a couple of times to look at some of their finds. Professor Windle was kind enough to ask me to come by."

"What is Professor Windle like?"

"Oh, rather bookish and introverted, exactly what you might expect."

"I'll meet him and the others tomorrow, I suppose."

"You may see some of his crew in here tonight. They sometimes come in for the evening when they aren't scheduled to dive the next day. This place gets very lively around eleven o'clock." He glanced briefly around the room as though to see whether any of the divers had shown up yet.

She waited until he looked at her directly again before saying, "I've heard that there was an accident here about ten days ago. A girl was killed." His eyes, on hers, didn't falter at all. "Were you here when it happened?"

"Yes. It was an accident involving the decompression chamber, according to the police." They had both lowered their voices. It was a serious subject.

"According to the police? Are you implying that it wasn't an accident?"

He raised an eyebrow at that. "Not at all. It's unfortunate, but—"

She knew what he was going to say, that diving accidents happened. Judy herself had once mentioned a fatal accident on another underwater project, but that one hadn't been anything like this.

But he didn't finish the thought as she expected. Instead he said, "Tell me, why are you interested in this girl?"

"Because—" She hesitated. She still wasn't sure she wanted to tell him the whole story.

"Because Judy Hamilton was your sister?"

She set her cup down so quickly some of the cream slid down the side. "Why did you say that?"

He sighed. "Look at it objectively, Ann. The newspapers here said Judy Hamilton had a sister in New York and parents living in Florida. Less than two weeks later a very attractive young woman, recently divorced, who says she isn't an archaeologist, comes to Tellin asking about the archaeologists. Asking specifically about the accident."

Yes, it was obvious. "What does 'attractive' have to do with it?" she asked dully.

"Because if you were homely I might be able to believe you had nothing better to do than come to an out-of-the-way village asking after strangers."

"Alex, my sister is dead and I don't understand how it could have happened. I intend to find out, if I can. And I certainly don't have anything better to do than that."

He studied her. "You aren't convinced that her death was an accident?"

"I don't know. That's what bothers me. I don't know what to think. It happened at such an odd time. She was having a dispute with one of the other divers—"

He seemed interested in that. "Which one?"

"I don't know." How could she make him understand that she knew Judy so well she couldn't help reading between the lines of her letters. What she had read there was an uneasiness, an uncertainty about her attitude toward the Omega project. Something besides the disagreement with the careless diver had been bothering her. Whether it was related to her death, she wasn't sure.

She considered for a moment before she took the letters out of her purse and handed them over to him. She forgot about the reference to Don until Alex had started to read, but she decided it didn't matter.

The *Gasthaus* was beginning to fill up. A group of young Austrians came in, and several more middle-aged men.

He finally looked up and said quietly, "It isn't much to go on, though, is it?"

Don't humor me, she thought. "I realize it's not very specific, but it wasn't like Judy to sound so discouraged. She loved what she did, she was always enthusiastic about it. It must have taken a lot to make her think of leaving here. If the others were really that inexperienced—" She was thinking it might have been negligence instead of a mechanical failure.

"So you have come to find out—"

"Whatever I can. If it was an accident I want to know it. But someone I talked to about it before I left New York, someone who ought to know, said it was very unusual for a winch to fail like that." She was referring to an editor where she worked who had written a book about diving. She had spent over an hour in his office talking to him about it. "It's just very rare."

"That's quite true. But rarely is not the same as never. All machines are fallible, just like people."

"All right, but there's something else. According to what I've been able to find out, if the decompression chamber was only ten feet below the surface where it was supposed to be, Judy probably wouldn't have been killed instantly the way she was. The difference in the air pressure between there and the surface wouldn't have been great enough."

"Not necessarily," he said grimly. "People have died from coming up too fast in swimming pools. If they get rattled and hold their breaths, the pressure doesn't equalize and the lungs—" He saw the look on her face and stopped short.

He reached over and put his hand over hers. "Look, Ann. Will you please let me give you some good advice?"

She met his eyes.

"Forget about this," he said. "Go home. There's no way you can bring your sister back."

She glanced away. "That's a cliché, Alex," she said. And it had been too close to what her parents believed. They thought she had gotten this idea about Judy's death because she was still upset about her divorce and was unconsciously looking for a way to escape into something else, a futile trip to Austria. But it wasn't true. Her father had told her, *It's something we will just have to accept. Honey, Judy is gone.* But she couldn't accept it until she understood it.

"Or better yet," he went on earnestly, "stay on, but forget about pursuing this. Let me show you around the Tirol. I won't be able to leave Tellin for a while, I'm not sure how long, but afterward—"

"No," she said flatly. Why was he so interested in what she did?

"Isn't there anything I can say to persuade you?"

"You could begin," she said slowly, deliberately, "by telling me the truth." When his expression didn't change, she went on, "You're acting as though . . . you're trying to warn me about something, and I don't understand what it is."

He leaned back while his eyes went over her face thoughtfully for a long moment. "All right then. Finish your coffee and we'll go for a walk. I'll tell you how it is."

FIVE

They went a short distance down the street until they were alone and sat down on a bench near the village fountain. There were lights in windows up and down the street but none so close that someone might overhear.

Alex lit a cigarette, let out a breath, and said, "I'm not going to ask you to promise not to repeat this. When I get through you'll understand why it's necessary not to talk about it, and if I thought you would, I wouldn't be telling you.

"It started a very long time ago, during the last days of World War II. The Nazis intended to make their final stand here in western Austria. Goebbels called it *Alpenfestung*, the Alpine fortress. They intended to fight on, guerrilla style, and if the Allies hadn't advanced so quickly, some of the most fanatical soldiers might have managed to hold on for many years. As early as the end of 1944 the SS began sending their families here—some people say the population of the Tirol doubled between 1944 and 1946. They also sent vast amounts of gold and other valuables. Part of it came out of the treasuries of occupied countries, the rest was expropriated from private citizens.

"After the war some of this treasure turned up in very strange places. Did you know that one of

General Patton's men discovered several million dollars' worth of gold and art works in a salt mine near Salzburg? And here in the Tirol gold bricks were camouflaged as roofing material. The French Army recovered that particular cache when a roof collapsed from the weight. No one knows exactly how much treasure the Nazis brought here, but it's safe to say that a good deal of it has never been found. Some of it probably never will be.

"Have you ever heard of a German general named Heinrich Krueger?"

The name sounded familiar but she wasn't sure, so she shook her head.

He took a drag from the cigarette and exhaled. "During the closing days of the war Krueger was in Tellin, up there at the castle. Just before the American troops got here he disappeared. Some of the villagers remembered hearing a small plane take off one night from that meadow at the end of town. The authorities assumed that if he made it to Spain or North Africa, he probably escaped to South America.

"When the Americans arrived, they captured a German soldier in Krueger's old headquarters at Innsbruck who reported he had helped load a truck with heavy boxes taken out of a vault. The boxes contained gold that had been diverted to Innsbruck when the Allies began sweeping across southern Germany. The truck went to Tellin with Krueger, but the gold was never found. Krueger couldn't have taken it all with him, because the entire load was estimated to weigh around a quarter of a ton."

"That's all very interesting, Alex, but what does—"

"Wait. You'll see. In 1965 the Austrian secret service received a tip that Krueger was alive and living in Argentina. Along with this information came a recent photograph of a man who did indeed look exactly like Krueger, twenty years older than when he disappeared. It had evidently been taken on a street corner without Krueger's knowledge. The Austrian officials thought it was interesting, but since there's no extradition between the two countries—Krueger is still wanted for war crimes, incidentally—they filed the information away. Then about two months ago they got another tip from the same source. It said that Krueger was planning to use the archaeology project here as a cover to retrieve the gold."

She had kept her eyes riveted on his face while he talked. "How do you know all this? Are you a policeman?"

"I'm what I told you I was, a biologist. But I have a friend who works for the secret service in Vienna—it's the counterpart of your FBI, I suppose. His name is Kurt Reinhardt. He knew that I spend my summers in Tellin and that I know the people here. He came to see me several weeks ago and asked me to help them."

"Help them? How?"

"By keeping an eye on what was going on. Watching for anything unusual that might develop. That's all. I had some reservations, a lot of them in fact, but none seemed serious enough to let me say no."

"But where did these messages come from?"

He threw the cigarette down and stepped on it, then leaned back with his arms resting on the back of the bench. He had been looking straight ahead as he talked, concentrating on what he was saying. "Anonymous letters, both times. They were mailed from Zurich, but that doesn't mean much. They could have come from anywhere ultimately. There are several theories about who might have sent them. Krueger has probably made enemies since he left Austria, he was the type who would. So it may have been someone who found out about his past and wanted to get even for some reason.

"It also happens that Krueger left a wife and young son in Germany in 1945, and they're probably still alive. It's possible, I suppose, that one or the other of them resented the fact that he never sent for them but simply left them in occupied Germany to fend for themselves. However, it isn't likely they would know about his plans at this point.

"The third possibility is more complicated. During the 1950s the Austrian Communist Party built itself an office building, a big ugly affair very much in the Stalinist tradition. It cost several hundred thousand dollars, and the party was very evasive about where it got all that money. They may have been sensitive because they knew the money came directly from Russia, but a newspaper reporter noticed that one of the prominent members of the party at that time was none other than the soldier who had reported loading the truck headed for Tellin. There was speculation that he may have diverted some of the missing gold and that it wound up in the party coffers. Actually he couldn't have done that, since there were other witnesses who verified that the truck arrived here with its cargo intact. But just the same the party was embarrassed by the rumors, because some of the gold must have come from the victims of the death camps.

"Anyway, one theory about the anonymous letters is that the Russian NKVD located Krueger sometime before 1965 and kept watch on him, then passed the word along when they found out what his plans were because they wanted to clear up the matter of where that gold went. They're like the Nazis that way, they bear grudges for a long time."

He shrugged. "It's even possible the Israelis may be involved. They don't forget either, for different reasons. A lot of people would like to see Krueger caught, or at least prevent him from getting away with that gold."

She was feeling more and more uneasy. "Why are you telling me all this? In order to frighten me?"

He turned to look at her. "Frankly, yes. I want you to be aware of the situation here, at least. No one is sure yet exactly what is going on. That's why it's so dangerous."

She was still reluctant to believe it. "I don't see how you can be sure these anonymous messages are true. It happened such a long time ago. This Krueger would have to be an old man by now."

"Yes, he would."

"Why did he wait so long? Why wouldn't he have come back for that gold years ago, if it's really here?"

"I've wondered about that too, but this may have been his first opportunity. In the past other people have tried to recover Nazi treasure in Austria. At least one man has died falling from a cliff while apparently trying to reach a mountain cave, for instance, and one or two others have drowned in lakes. But those attempts occurred in unpopulated areas where anyone could come and go without being seen. That's not the case here. If the gold is somewhere in Tellinsee—in the Tellin lake—the diving expedition would seem to be a perfect cover for an effort to retrieve it. And it's probably the first chance that's come along for him."

"But an old man like that . . . to come back now?"

"No, it wouldn't be possible for him to do it by himself. He would have to send someone younger to bring the gold out of the lake. Or so my friend Kurt believes."

"You think someone Krueger sent here may be at the castle right now, with Professor Windle's group? Is that what you're saying?" She looked down the street. The castle had merged into the shadow of the mountain behind it, forming a gigantic black wall. What he had told her would change things, but she wasn't sure how. "Is it possible that my sister discovered something about this and . . ."

"Don't jump to conclusions," he said sternly. "The police didn't find any evidence that the winch had been tampered with. They were completely convinced it was entirely accidental."

"*Were?*"

He said carefully, "Ann, as far as I know, it was."

She sensed that he was holding back again, perhaps afraid of putting ideas in her head that would lead her off on the wrong track. "You don't have any idea who it might be? Which one?"

"No, it could be any of them. There are three Latin American boys and a girl."

That would be the Phillips girl Judy had mentioned. "But Karen Phillips is from Florida, isn't she? How could she be mixed up with your General Krueger?"

"They aren't ruling her out, even though the Latin Americans are higher on the list, naturally."

"How do they expect to find out which one it is?"

"The police are making inquiries, very quietly of course, about everybody in the expedition. The problem is that none of the divers is from Argentina, according to what they claim. None of them has ever been there, so far as the police have been able to find out. And it isn't likely that Krueger would have gone to another country to enlist a stranger. He would want to use someone he knew well, someone he could trust. So the theory is that one of Windle's divers is an Argentinean using a false identity."

"How would that be possible? Judy had to send

in an application and references. And she sent her college transcripts too, I'm sure."

"It isn't difficult, really. One way is to borrow the identity of someone else. It doesn't take a great deal of trouble. You could, for example, find out the name of someone your age who was a former student at a large university and simply write for references and transcripts using that name. Barring the very unlikely possibility that the registrar knows that student personally and recognizes the phony address, the material would be sent off without any question."

It had never occurred to her that something like that was possible, but she realized that it could be done. Even if the university had another address listed for that student, they would probably merely assume that he had moved.

He said, "The police in other countries have been cooperating by making inquiries at the universities Windle's divers come from. They are checking the old addresses, when they can find them, to make sure there isn't another person with the same name back there. So far they haven't come up with any doubles. If Krueger planned this wisely, he may have picked the name of a person who wouldn't be easy to locate, someone who had left the country, for instance, and had no family."

"It's too bad your informant couldn't have found out who Krueger was planning to send."

"Yes, it would have made things much easier."

Something occurred to her. "They investigated my sister too?"

"Yes, even your sister." He was frowning. She realized what the tenseness came from. This assignment he had taken on was bothering him. "Kurt did tell me she was cleared, because she had never left the United States before. They must have found that out by checking with the passport office. And there was a picture of her in her university yearbook that matched."

It didn't sit well with her that they had investigated her sister. She was trying to put it all together. "From what you've said, the police can't be sure Krueger really has sent someone here, if they haven't found out anything, well, *suspicious* about any of the divers."

"That's true. But they're still looking."

"What happens in the meantime?"

"We have to watch and wait. They don't want Krueger's agent to be frightened off before he recovers the gold and starts to bring it out. That's probably the only chance there is of ever finding it. And possibly he can lead the secret service to Krueger, if Krueger has made the mistake of coming back to Europe to supervise this operation at close hand."

She thought for a moment about what he had said. "So that's why you don't want me to go up to Eagle's Castle, I suppose. You're afraid I might do something to upset the plan."

"That's part of it, of course. Whatever is going to happen should happen soon, because the expedition will be finishing up. But mostly I don't think it's a good idea for you to go asking a lot of questions about your sister. You might put yourself in danger."

She realized that what he said made good sense, and yet she couldn't accept it.

"But if what happened to Judy was an accident as you say, then it doesn't have anything to do with Krueger and what you're talking about. I don't see how my questions could do any harm, if there's no connection. Besides—" It all hinged on those two anonymous letters. That's all the evidence the police had that Krueger was using the Omega project, and it was questionable evidence. Their informant could be mistaken, or lying. There were a lot of cranks in the world.

"Ann, please don't get involved in this," he said urgently.

"You don't seem to understand," she said impatiently. "I *am* involved. My sister was very special to me. I can't turn around and go home now after coming all this way without at least talking to the people at the castle who were there when it happened. Especially not on the basis of an anonymous letter that may or may not be true. Don't you see?" She stood up and faced him squarely.

"I see that you are a very stubborn woman," he said.

She was irritated by that. "Oh? And I suppose if I were a man you'd say I was 'determined' and 'strong-minded.' " Immediately she wished she hadn't snapped at him.

Instead of getting angry, he laughed. "So you're

a feminist besides." His laughter broke the tension for both of them.

"No, I'm not, really. I'm just tired, I guess. I'm sorry, but I don't think I can do what you ask, Alex. And I think I'd like to go back to the Bruner's now, please."

At the front door she relented a little. "Anyway, I'm grateful that you told me, Alex. I just feel that I have to talk to Professor Windle and the others." The story he had told her was an added complication, but she couldn't think of it as more than that. Still, she said, "I won't let anyone know what you've told me, I promise."

He still looked worried. "Think it over tonight before you make up your mind definitely, will you? If you should need me for any reason while you're here, I'm staying at the Müller cottage down by the lake. Anyone in the village can tell you where it is."

"Thanks, Alex. I'll remember."

He realized from the sound of her voice that she wasn't planning to change her mind. "If you insist on going to the castle tomorrow, I'll pick you up there for lunch, all right?"

"Yes."

He frowned. "No, wait, that won't do. I just remembered I have to meet Kurt in Innsbruck tomorrow. He wants to talk with me about this business. It would have to be the day after. You'll still be here?"

"Yes, that's fine." She had intended to stay about a week, although now she wasn't sure.

"I wish you would agree to continue staying here at the Bruners' and not—"

"I can't promise, Alex. Really. But I'll be careful." He saw that it was no use. There was nothing more he could do to stop her.

Finally he kissed her good night, and not too briefly either. After going inside she realized she had liked it more than she had wanted to.

As for Alex, he left feeling guilty about not being able to tell her the whole truth, the part that involved him in a deception.

SIX

She got up early, pulled on a navy pantsuit and jersey blouse. She put her nightgown and toothbrush into the suitcase and brought it downstairs with her. During breakfast in the kitchen she tried out her scanty German with Frau Bruner, who had already eaten and was busily cleaning the top of the iron stove. Frau Bruner's German turned out to be quite different from the kind her grandmother had spoken, and Ann had difficulty understanding her.

She tried to turn the conversation to the archaeology project, but the older woman had little to say except that they were "looking for old things in the lake." And the baron? She made a cryptic remark about "the war" and added with a sigh, "*Er ist nicht derselbe.*" (He is not the same.)

As for Schloss Adler, Frau Bruner apparently regarded it as the most commonplace thing in the world to live in the shadow of a sixteenth-century castle. She dismissed it with a shrug and the comment that it was "*sehr alt*," very old. She attacked a spot of grease on the stove as though it were an old enemy. "Let the past remain in the past is my way of thinking," she said gruffly, and that ended that.

Her husband came in, a tall gaunt man with receding gray hair and a long face with intelligent brown eyes and a thick mustache. His skin was like tanned leather except for his rosy cheeks. He greeted her warmly and she liked him immediately. His name was Max. She was relieved that he spoke English.

"I think I saw you with my friend Alex at the *Gasthaus* last night," he said.

"Were you there? I'm sorry we didn't see you."

"Oh, you were just leaving when I was coming down the street, or I would have spoken to you." He got a cup of coffee from the pot on the stove and sat down across the table from her. "When you see Alex again, tell him we have a chess game to finish."

"Yes, I will."

He asked her where she was from. When she told him New York, he asked her about New England. "I've heard that New Englanders are something like us Tirolese," he said. "Very conservative. Frugal and hard-working too."

"That could be," she said. "New Englanders are supposed to be like that. Vermonters, in particular."

"I wonder if it might have something to do with the geography, you know. Do you suppose the mountains toughen a people's character, make them hard and unyielding, like those Alps out there?" Gesturing toward the window, he asked the question as though he really wanted her opinion, and she liked that.

"Wherever making a living is hard, I suppose people have to be tough and self-reliant. In the tropics they can afford to be more easygoing and more—" She searched for a word. "More wasteful?"

He nodded enthusiastically. "Yes, that's what I had in mind exactly. Many years ago I met an American G.I. from your state of Alabama. It was during the occupation just after the war. What an easygoing fellow he was! Not a care in the world, it seemed. He was amazed that people could live here in these mountains with so little farm land and such a short growing season. He told me how it was in the American South, with all that flat rich land. He made it sound as though the farmers had only to throw seeds on the earth and stand back to watch things grow."

Now that she thought of it the Tirol did remind her of New England. The villages in both places seemed neat and proud, with a healthy respect for tradition.

"Even our economies are the same, I understand," Max went on. "Once we Tirolese relied on our dairy cattle and a little farming and hunting. Now it's tourists and skiing." He waved his hand in the air as though he didn't like the change. She had noticed a deer head and a rifle on a wall in another room when she came down the stairs. Max had evidently done some hunting himself.

His wife said something in German, and he translated. "She wants to know if you'll be staying over tonight. You are very welcome, but she would like to know how to plan the meals."

"Oh, I don't think I'll be coming back, thank you. I'm going to drive up to the castle today and I hope they'll have room for me there. That's where I was heading yesterday, but Alex suggested I say here instead since it was late."

She was surprised when she saw the couple exchange a brief look. There was something like alarm from the woman and a silent reprimand from Max in return. All very quick, but she was certain she hadn't imagined it.

"So you may not be back then?" Max asked in a friendly way.

"I'm not sure. If I'm not back by early afternoon, let's say, you shouldn't expect me."

"That will be fine. We hope you'll return, but if not, be careful," he said. "On the road up, I mean. It is quite narrow and steep."

In the morning light Schloss Adler looked somewhat less formidable than it had the day before. Her car was waiting in front of the house just as

Alex had promised. The road went straight through a meadow and curved gently to cross a wooden bridge. Down at the stream on her right there was a temporary camping area with a few tents. Young Austrians on vacation, probably. In a field on the other side of the road two young girls and a woman in a dark dress and kerchief were gathering light brown hay. Using long-handled wooden rakes, they pitched it up and over low hayracks that ran like a fence down toward the lake.

The road forked. One branch curved away to the right into the mountains and the other, a narrow lane, went toward the castle. She slowed down and turned left. It was a twisting, difficult drive. Evidently the road had started out as a footpath and had been widened as a concession to the automotive age; at times it was barely wide enough for her small car. She rounded one hairpin switchback and then another as the road struggled upward.

Abruptly the drive leveled off. She passed through an opening in a thick stone wall and found herself on the castle grounds. The castle walls had crumbled in places. Stones had fallen away and scattered onto the edge of a broad courtyard. Grass grew sparsely around several well-worn paths. In the center was a concrete base where a fountain must have been at one time. On either side a massive gray building rose up to a slate-colored roof. She followed the gravel lane to what looked like a parking area by the door of the building on her left, and stopped. On the ground floor of the other building an arcade led to the round tower overlooking the lake. A blue van truck was parked near it, next to another crumbling wall. When she stopped the motor there was no other sound.

She got out. From the corner of her eye she caught the blur of a movement behind the window near the castle's large front door. But when she turned there was no one there. She decided that it must have been the shadow of someone passing by the entrance. She walked over to the door and rapped with the brass knocker. After a long moment the door opened.

The thin hawk-nosed man standing in front of her was only a few inches taller than she was. Though he must have been in his sixties he stood stiffly erect, his gray hair cut short like a soldier's. He regarded her with his stern, sad eyes. "*Guten Tag, gnädige Frau?*"

It was definitely a question: Good day, madam? She realized that she was being subtly kept at bay and that if she didn't cite a very good reason for being there she wouldn't get past the front steps. Summoning up her German, she explained that she had come to speak with Professor Windle and that she would also like to rent a room, if one was available. For a few days, a week or so at most.

He instructed her to wait and disappeared, closing the big door. She waited. Another door slammed behind her, making her jump. The metallic sound had come from the van across the courtyard. A dark muscular young man in swimming trunks had just closed the rear door and was standing there cradling a scuba tank in his brown arms, looking at her. Even from that far away she could see that he was smiling at her. They stared at each other for a long moment. Then the front door reopened behind her.

The stiff little man was back, directing her to follow him. As she stepped inside, she became aware of several things in rapid succession: the thickness of the outer walls—six feet, at least; a dimly lit foyer with a faded oriental rug, an antique table, and straight-backed chair; a smell of old mortar and candle wax. She followed down a hallway leading off to her right and through open double doors.

A woman was sitting in a high-backed chair facing the doorway, too much like a member of royalty granting an audience to Ann's mind. The woman was perhaps fifty-five. Her gray hair still held a tinge of reddish brown and was pulled back in a tight bun low on her neck. Her nose was thin at the bridge, her eyes dark and deep-set. She wore a jeweled pin at the collar of her blouse. A small woman, her voice was surprisingly forceful as she greeted Ann and asked her to sit down.

"I am Olga Baroness von Toblen," she said, very precisely. "Wilfred tells me that you speak luffly German, but I hope you vill not mind if I practice my English. You've met my husband Wilfred, I see."

Ann drew in her breath. She had assumed the morose old man was a servant. "Yes." She glanced back and saw him taking a seat nearer the door.

The room was rather depressing. Dark wainscoting on the lower walls matched the wooden ceiling, with white plaster in between. Several hairline cracks in the plaster, a few worn spots on the upholstery on the heavy furniture detracted from everything else.

"You are from America, Miss Cole?"

"Yes. New York City."

"And you are a friend of the professor here, I belief?"

"No, I've never met him. But I wanted to talk with him. That's why I came."

"You vish to stay for a time here?"

Ann said that she did.

"Our guests usually stay in the *Unterschloss*, the other part of the castle. The only restored rooms are at the moment occupied by the scientists who are staying here with us, but possibly"—she looked pointedly at her husband—"the upstairs bedroom. Do you think so, Wilfred?"

She turned to see him nod his agreement, still looking glum.

And so it was settled. The two women exchanged some conventional remarks about the weather and about airplane travel, and then the baroness told her she could probably find the professor in the tower. Ann stood up.

She turned to include the baron in her farewell, but the sad-eyed man had already slipped out of the room.

SEVEN

The courtyard was empty, the young man she had seen earlier nowhere in sight. She found Windle on the ground floor of the tower. It was a huge circular room with deep windows and stone walls, and she noticed first of all how cool and humid it was. There was a clutter of odd-looking equipment on the floor along one side, and several large worktables covered with an assortment of artifacts and charts. She saw him at a table in the back. He was bent over, peering at something with a magnifying glass. In his fifties, he had short brown hair that was sticking up in places. He was wearing a khaki shirt open at the collar.

He heard her walking across the room and glanced up. His glasses made his eyes look big, giving him a startled look. He motioned her toward him. "Here. Come look at this."

When she stood next to him he jabbed a finger at the large photograph he had been studying. He held the magnifying glass for her. "Tell me, is that a handle of a jug or isn't it? See, that curved line there."

The picture showed the muddy lake bed with rocks the same color brown as the mud and what might have been a curved jug handle. Over all of it there was a man-made grid of some kind, evidently put in place so that distances could be gauged in the photographs. There was a small plaque on one of the metal rods that said: "Sec. 12-B."

"I can't be certain," she said.

He sighed unhappily. "That's the trouble with getting old. The heart goes, and you have to depend on other people to do the diving for you, then you're never quite sure they're doing it right. The eyesight isn't what it used to be either, so photos don't help. Who are you, anyway?"

"Ann Cole," she told him.

He cocked his head, eying her. "Do I know you?"

"You knew my sister. Judy Hamilton."

He recoiled a little in surprise. "Oh my, I'm so sorry about that, you know. But what are you doing here?"

"I came to talk to you, actually."

"About Judy, I suppose?" He twirled the glass in his hand nervously, then put it down. "I don't know what I can tell you. The police said they had written you and your parents giving all the details of how it occurred. I should have written you too, I know. Somehow I just couldn't bring myself to do it. Believe me, I'm very sorry. I've been involved in nine expeditions, and this was the first serious accident. We use the very best equipment and still something like this—there was no way to foresee it."

She tried to put him at ease. "I'm not blaming you for what happened, Professor Windle. I just want to know—you see, the explanation the police gave wasn't clear to me. Would you mind if I asked you a few questions about it?"

"Certainly I don't object." He lowered his voice. "I'm afraid you're only making it harder on yourself, though, if you don't mind my saying so."

"Don't worry about that. It would be much worse if I'm left with my questions unanswered."

He got up slowly, collecting himself. He was taller than she had expected. "I was just about to leave for the boat," he said. "If you want to come with me I can show you—how it happened."

They walked behind the blue van to the edge of the courtyard and down an extremely steep path to a rocky beach where there was a dinghy half out of the water. He helped her in, pushed off, and started the electric motor. The diving boat was about a hundred and fifty yards offshore. Approximately sixty feet long, it resembled a converted barge riding low in the water. When they pulled alongside she grabbed the ladder and climbed up the rungs. The young man she had seen earlier materialized from somewhere on board and helped her up to the deck. He was wearing a wet suit now and she noticed he had a knife strapped to his calf. Standard diving gear, she reminded herself.

After the professor followed her on board he introduced them. "This is Martin Cabrera from the University of Mexico City. Ann Cole."

"I am very pleased to meet you," Martin said.

He was smiling at her with what was either genuine interest or old-fashioned gallantry. He was obviously Latin, with smooth dark skin. His soft brown eyes were as pretty as a girl's. Only his nose, slightly flattened at the bridge—a boxer's nose, she thought—saved him from being too handsome. He was a head taller than Ann and perhaps a year or two younger.

"Are you a new diver come to join us?" he asked. There was a flash of white teeth as he added, "I hope so."

Windle explained who she was and Martin's mobile face turned instantly sympathetic. "I am sorry," he said quietly. "It was a tragedy. We miss your sister very much."

She was introduced to the others on board. They found Karen Phillips near the stern, stooped down beside the compressors and air tanks. She had long straight brown hair with a few lightened streaks, parted in the middle. She was wearing big round sunglasses with wire frames. The Gloria Steinem look. Ann saw that she was tinkering with a small motor that had its innards spread out on an oily rag on the deck. She greeted Ann with a cool "Hi" and went back to her work.

The other two divers had just come out of the water. Ann had trouble distinguishing between them. Both were short and stocky, in their early twenties. They glanced at her from time to time without smiling as they shucked off their wet suits.

Ann remembered reading somewhere that some Latin Americans claimed to be able to tell which country a fellow Latin came from just by his appearance, from the shape of his face and the texture of his skin. Yet these two were from different countries, and they seemed alike enough to be brothers. Eduardo Alvarez was from Chile, Jorge Domingo from Panama. Eddie and George, Windle called them. Neither spoke English, apparently. She did notice some difference between them—George's round face was slightly broader than Eddie's. And his brown eyes were more evasive.

The stern was crowded with equipment. In addition to the two large air compressors there was a chest-high rack holding diving gear and tools and a row of air tanks. The cabin near midship was small with long open windows on two sides. As they walked forward Professor Windle showed her where the air line to the decompression chamber went down off the starboard side. There was another hose beside it that he said sent air to the bottom of the lake. "The air's used to clear off silt," he explained, "and to fill up the balloon lift when they want to send something to the surface."

The winch was on the starboard bow behind another rack—this one held wet suits, flippers, and goggles—and a large steel contraption that the professor told her was a "one-man recompression cylinder," for emergency use should any of the divers show signs of the bends. They had never had to use it, he told her. Behind it, along the front of the bow, she saw that there was a clear passage to the other side of the boat, though it was not as wide as the clear deck area in front of the cabin.

A steel cable an inch thick ran out of the winch, through an eye hook at the edge of the deck, and down into the water like a long gray tongue. A casual look around showed her that anyone standing at the winch where they were couldn't be seen from the cabin or the other side of the bow, let alone from the stern, where the cabin blocked the view. She leaned over the rail and glanced at the water four feet below, then turned back to the winch.

"How does it work?" she asked him.

"You see, when this is down"—he pointed to a long arm on the top—"that holds the cable in place. To release it, it must be lifted up and pulled back." He went through the motions in pantomime, not touching the arm. "It was supposed to be completely safe, but something went wrong, something apparently slipped or a cog broke and the thing flew back. There is a tremendous pressure being exerted from the air in the chamber down there. . . ." His voice faded in uncertainty.

"And then what happens?" she prodded him.

"The cable begins to feed out very rapidly, going down to the bottom first and then up to the decompression chamber. As the chamber rises, the air inside it expands and it goes even faster. It shoots up very quickly."

"Until it comes to the surface."

"Yes, then the cable stops feeding out, of course, because there's nothing pulling it any longer."

She looked at the ugly machine. "You said

something apparently slipped, or broke. Surely the winch was examined afterward to see what went wrong?"

"The machine was in too bad shape afterward, I'm afraid. You see, as soon as Martin heard the cable spinning out, he ran immediately—he was already on his way back here to put his wet suit in the rack over there. He said he had just rounded the front of the cabin on the other side of the boat when he heard it. He ran over here instantly and slammed the arm back down trying to stop it, but unfortunately it was already too late. It was all over so quickly. And though Martin had the best of intentions, well, I don't know much about machinery, but it seems that when he bore down on it, some of the cogs were sheared off or something. In any case, the police contended it was useless to try to figure out what happened after that. Actually we had to have the machine completely overhauled."

Now that he had started to talk about it, Ann noticed, he couldn't seem to stop. "It's supposed to be in fail-safe condition now, but frankly I still don't trust it. I've taken the precaution of having the divers sit in the open hatch when they use the chamber now, two at a time. With their feet in the water they could feel any sudden upward movement. Not that we expect anything like this again."

"There's something else I don't understand," Ann said. "My sister told me that she always dived with a partner. Yet she was alone in the decompression chamber that afternoon. If she went into the water with someone else, why didn't they come up at the same time?"

Karen Phillips had just come over to stand near the professor's elbow. "I think I can answer that," she said, drawling the words out. "It was one of those things. Judy was diving with Eddie that afternoon, and they went down first. Martin and I went in about a half hour later. Not long after that, ten minutes or so, Eddie had some trouble with his regulator and decided to go up. The rest of us worked down there for a while, and when Judy's time ran out all three of us went up to the chamber together." She stopped there, as though the rest explained itself.

"But she was alone," Ann said. "I still don't understand."

The girl shrugged as though she were dealing with an idiot and tossed the long hair back with a jerk of her head. "See, it's like this. The longer you stay down, the longer the decomp time takes. Like if you dive to a hundred feet for fifty minutes you have to take three or four minutes getting up to the chamber and then stay there for twenty-five minutes. That's according to the Navy's chart, the standard chart we go by. Except we always add extra time to what the chart says as an added safety factor. Martin and I had been down a much shorter time that afternoon than your sister, so we could go up to the boat long before she could."

Ann studied her face. The nut-brown eyes were an elusive blank. "Do you remember what happened after that? After you and Martin got back to the boat?"

"Sure. I took off the wet suit and got out a towel and lay down over there on the bow deck to get some sun." She motioned in the direction of the opposite side of the boat. The place she meant was out of sight behind the machinery.

"I couldn't see the winch from there, of course," she went on. "I didn't know anything was wrong until I heard Martin running across the deck, around the front of the cabin. Then I heard that terrible grinding of metal when he threw the arm of the winch down. And then, at the same time almost, the decomp chamber burst out of the water right over there." She pointed. "I heard it when it came up, even though I couldn't see it."

"Where were the others then, Eddie and"—she had to search her mind for the name—"George?"

"Back on the stern," the professor said. "I remember the police asked all of us these same questions. Yes, they were back there stowing away some of the gear, I believe. George was the first one in the water. He jumped in and towed the chamber over to the boat."

Evidently no one was near the winch when it failed. George and Eddie were farthest away, with Martin and Karen both on the bow, but on the wrong side. She turned back to Karen. "Did you say you saw Martin run over here to the winch?"

"No-ooo," she said carefully, twirling a strand of her hair around a forefinger and frowning off over Ann's shoulder. "I *heard* him. Running. I guess I could hear it because I had my head on the deck. By the time I raised up on one elbow to look he was already over there."

"He apparently heard the cable reeling out," Ann said thoughtfully, "but you didn't, I take it?"

She shook her head. "No, the compressors were running that afternoon. They make an awful racket."

"That's right," the professor said. "I remember that distinctly, the noise the compressors were making. I have a very good memory for sounds, strangely enough," he added dreamily. "Even sounds from my childhood. I can still recall very clearly certain sounds that I heard when I was quite young."

Karen threw Ann a look that said, *Windle is a bit of an oddball, but we humor him.*

Ann wondered now about the other two divers. Professor Windle had implied they had been together at the time. But the stern was roomy. They might have been separated, and one of them might have come forward to the winch along the starboard rail, avoiding passing by Martin on the other side of the boat. Without even thinking about it, she was already entertaining the possibility that someone might have disengaged the winch deliberately. But she was trying to cover all the possibilities.

"Had the winch ever shown any signs of trouble before? Had it ever been worked on?"

"No, not at all," Windle said emphatically. "It was new. So was the decompression chamber, for that matter—I've never used one on an expedition before."

"I don't believe you told me," Ann said, "where you happened to be at the time, Professor Windle."

"Why, I was in the cabin. Writing a letter to my wife. So you see, none of us was, unfortunately, near enough the winch to help matters. Probably once the machine failed, nothing could have been done in any case."

Karen said, "The accident has changed everything around here, I can tell you that. We used to laugh and kid around a lot. But now—" She shrugged.

"That's true. Very true," Windle agreed sadly.

Ann and Windle went inside the cabin, leaving Karen to go back to the stern. The cabin was about eight feet square with steps in the center going below. There was a desklike shelf built into the front wall with a swivel chair facing forward. Sitting there he couldn't have been able to see what was going on outside without pushing back from the desk and twisting around to look out one of the side windows.

"Do you remember seeing Martin standing out there?" she asked him, turning toward the port window.

"Mmmm, no. Actually I wasn't paying much attention at the time. There was no reason for me to, really. Martin had finished his work for the day. They were all finishing up, and—" He half laughed nervously and threw one of his hands into the air. "Really, I'm not quite sure what you're driving at."

"I'm sorry. I'm only trying to visualize exactly how it was."

"It was terrible. That's all I can tell you. Simply awful. I always warn my divers never to come up rapidly except in an extreme emergency, and then they must remember to breathe out constantly to avoid possible lung damage. But your sister, being in that chamber, was caught off guard. Not knowing what was happening, she didn't have time to think what to do."

He sat down in the swivel chair and leaned back, closing his eyes as though he were listening to something. "I can hear it now. The dreadful shriek of that cable spinning out. At first I couldn't decide what it was. Then I ran outside. But it was too late. Even if Martin had gotten there sooner, I don't think it would have helped. Once it picked up speed, nothing could stop it." He opened his eyes.

She was trying to think. Something didn't appear to add up. "You heard the noise and— How long did it go on, do you think? From the time you first heard it until the chamber came to the surface?"

He frowned at the ceiling, pondering, "Oh, it was very rapid, even though the chamber is quite heavy and has a lot of inertia to it. Definitely under five seconds. Not much under though. Three or four, probably. Anything less than that and Martin wouldn't have had enough time to run over there before it stopped, you see. He was only a few feet away, but even so."

"Are you sure about the length of time?"

He closed his eyes again. "Now that I think back, I can hear it. That's a sound I'm not likely to forget, ever. Yes, I'm certain."

The editor Ann had talked to had emphasized how the chamber would have popped almost immediately to the surface. Three or four seconds, she knew, *sounded* like a shorter period of time than it actually was: a thousand one . . . a thousand two . . . a thousand three . . .

She asked him to go back to the winch with her. She went to the rail and said, "No one is down there now, is that right?"

He nodded.

"Then could you please throw the arm on the winch and let it come up? It wouldn't be too much trouble to take it back down again, would it?"

He looked puzzled but said, "No, it wouldn't. Ordinarily I would bring it up with the motor, but I suppose there's no harm." He was humoring her although he still didn't understand what she was getting at.

When he pulled the lever back she kept her eyes on the second hand of her watch. One . . . No sooner had the hand reached one than she heard the dull climactic splash, looked up and saw the bright yellow ball bobbing in the water. It had come up from ten feet in approximately one second.

The divers all heard the splash. Ann saw that the two Latin-American boys were at the rail near the other end of the boat, talking excitedly to one another. Karen was nearby, standing impassively with her hands on her hips. She took off her sunglasses and wiped them on her blouse. Martin, who had been talking with her, turned around with a quizzical look and now was coming toward them, probably to ask what was going on.

They didn't understand what it meant, perhaps, but there was no doubt that the professor did. He was staring at her, his eyes full of disbelief. "But the chamber *couldn't* have been deeper that day." He whispered it as though he were talking to himself. "Someone would have had to lower it."

She turned her back on all of them and stared out over the lake. It was a beautiful day. The bright yellow metal ball looked as cheerful as a daisy in the water. The sun shone down on the gray jagged mountains and the equally jagged conifers and on the mirror blue of the lake. The toy-sized village with its quaint white houses clustered around the pointed steeple was like something out of a child's picture book. Even the huge stolid castle looked tranquil and forgiving.

It didn't look at all like a setting for a murder.

EIGHT

Later that day she would talk to Windle again. The four divers were given the afternoon off and had taken the opportunity to drive into the village. She and Windle went into the tower and sat down facing one another across the worktable in the back of the room.

"I still believe there must be some mistake," he said unhappily. "Some logical explanation for the discrepancy in time."

"If you can think of one please let me know," she said dryly. She wished she could still believe it might have been an accident. Ever since that morning she had been trying to think of some other way to explain it. Could there possibly have been that much difference in time—only a few seconds, but a percentage difference of at least 300 per cent between one second and three or four—if the chamber had been at ten feet both times?

She didn't think so. Once the winch was released, the cable must have spun out at its maximum rate. How else could it be explained except that someone must have deliberately lowered the chamber to a more dangerous depth before releasing the winch?

He shook his head in exasperation. "It *couldn't* have been down that far that afternoon. It's *always* kept at ten feet. Besides, the others who were there with Judy earlier would have noticed it wasn't at the proper depth when they came up."

"I know." It meant that the chamber must have been lowered sometime after Karen and Martin left her, probably just before the arm was released. "Do you remember hearing the winch motor running? Sometime just before that, I mean.

Pulling the cable in instead of letting it out?" She knew the motor would have had to be used.

"No. But I wouldn't have. The motor is quiet compared to the compressors. But there was no *reason* for anyone to turn it on. Why on earth would anyone do that?" He was staring at her.

She could understand how he felt. She could hardly believe it herself. All along, it seemed, she had nursed the silent hope that the police were right. She had come there hoping that her doubts—that uneasy, ill-defined feeling that something was wrong—would prove to be unfounded. She hadn't suspected that it would be murder. *Someone murdered Judy.*

She realized now what Alex Schuler must have suspected and didn't want to tell her. The diver Judy was concerned about might have been going off to look for those boxes of gold hidden somewhere in the lake. If that diver knew Judy intended to report it and was afraid that it might mean being dismissed from the project and having to give up his "cover" . . .

"Professor Windle, there was something my sister mentioned in her last letter. Did she ever tell you that she was having a problem with one of the other divers?"

"Not that I recall, no." He looked genuinely bewildered.

"She wrote me that someone she was diving with kept wandering off when they were supposed to be together. She told me she was planning to tell you about it if it happened again. That's a serious infraction of diving rules, isn't it?"

"Certainly. I don't permit that kind of thing. It's a basic safety precaution to stay together. But she never said anything to me. You don't have any idea who it was?"

"She didn't give me the name. I suppose she thought a name wouldn't mean anything to me." And another reason occurred to her now. Two reasons, actually. If Judy had written the letter while she was on the boat, she might not have put the name down because she was afraid someone might walk by and happen to look over her shoulder and notice it. And knowing her sister, Ann realized that she probably hated "telling" on someone so much that she may have even subconsciously resisted putting the name down in black and white.

He raised his head and made a helpless gesture. "It's true that some of my divers are relatively inexperienced. That's one of the purposes of the expedition, to provide training for them. However, I didn't let them enter the water until I was certain they were drilled in the safety rules. There's no excuse for anything like that."

"Do you know who it might have been? Who did Judy usually dive with?"

"They rotated assignments. I couldn't tell you who she dived with on any given morning or afternoon because it was very informally worked out. It wasn't as though I posted an assignment sheet that I would save. It wasn't like that. If you want me to ask about it, I will. I'd like to know the answer myself."

She shook her head quickly. "No, please don't.

In fact, I'd appreciate it if you didn't say anything about any of this, Professor Windle." She had been grateful when, out on the boat, he had turned Martin's questions aside with vague comments about how she had "wanted to see the chamber come up." He hadn't said anything about the discrepancy in the time it took. He had probably been too astonished to put it into words.

He was still upset. "Even though they aren't an experienced crew, I can't believe any of them would be stupid enough to turn the winch on and then release the arm. What possible reason would they have for that?" She knew that he was now thinking what she had thought earlier, that it was a human error of some kind. Negligence. But then, he didn't know about Heinrich Krueger and the gold. He didn't know there was a possible motive for murder.

He said, "Perhaps someone was walking by and slipped and accidentally hit the on button that started the chamber down. And then, seeing what was happening, tried to stop it and accidentally threw the release arm back." But he already realized that it wasn't likely. It took a real effort to lift that arm up. And there was no reason for anyone to be over there by the winch. His imploring glance slid away from hers like melting butter.

"What are you going to do now?" he asked.

She had been thinking about that. This was obviously something the police should know about. But without an eyewitness or any other evidence, she didn't think they would get very far.

"I'd like to stay here for a while and talk to your divers, if you don't mind. Possibly I'll be able to learn something more."

"Of course, if you're sure that's best. I should probably make some inquiries of my own."

"I wish you wouldn't do that, Professor Windle. I really think it would seem more natural if the questions came from me."

"Yes, I see what you mean. If I were to take an interest in it now, it would be obvious I had discovered something amiss. Possibly the person responsible would be reluctant to talk to me about it, whereas you would naturally be interested in finding out all the details—"

"Yes, exactly." She hoped that the experiment of raising the chamber hadn't already given her away.

"Well. I suppose you have a right, certainly. If you find out anything of relevance, please let me know. And if you want me to talk to the police about this again, I will."

"Thank you. I was thinking of asking you to do that later on, after I've had a chance to see what I can find out."

He told her that he would keep her confidence in the meantime, and that if there was anything he could do . . . He got up to see her to the door. As they walked past one of the worktables he paused to show her some of the finds that had been brought up. There were jugs and pieces of bowls, ax heads and other tools, some with the wooden handles still preserved. Each one was tagged with a date and number. The pottery was encrusted with a thin layer of hardened tan sediment. It felt like stucco when she moved her hand over it.

"These will have to be carefully cleaned of course," he told her. "We haven't made any spectacular discoveries so far, but every new find adds to the general store of knowledge about these lake people. They were very ingenious. Look at this." He picked up a small odd-looking gadget.

It was a safety pin, shaped like a capital D with the pin in the downstroke. She felt the tip. It was still sharp. "Bronze," he said. She handed it back to him and he said, "Even after all these years I get a strange feeling whenever I handle these old things. I try to imagine who made them, what they looked like, the way they thought." He glanced at her with a self-effacing smile. "Not a very scholarly attitude, I suppose."

"I have the same reaction myself."

"These pins are quite common in Bronze Age sites, but this is the only one we've found here so far. Actually, your sister was the one who discovered it. She was very pleased, I remember."

He stared at the relics on the table. "I suppose I like dealing with dead civilizations because they're so much simpler and safer than our own. If it's true that your sister's death wasn't a simple accident it will probably create a scandal and the rest of the expedition will have to be scrapped. But that doesn't matter. I'll help you in every way I can."

She left him with the feeling that he was a very nice man.

"This expedition hasn't been as exciting as I had hoped."

Martin was standing at the tower battlement overlooking the lake, leaning with his fist against the stone wall. The battlement formed huge U-shaped notches all the way around the top of the tower. Through one of them he and Ann could see the lake and the mountains on the other side growing purple as the sun went down.

After leaving Professor Windle, Ann had spent over an hour walking around the grounds thinking. When the van came back from the village Martin had found her and invited her to go with him up the tower stairs to admire the view.

"My diving teacher told me about digs he'd been on that sounded much more interesting. At Chichén Itzá, for instance, they had to dive over 140 feet into a Maya cenote. That's a narrow well, you know. Yes, that was exciting. The Mayas used to throw sacrifices into it. Can you believe it, they brought up over 30,000 artifacts out of that one well. Most of them were interesting too, because the Mayas sacrificed their most valuable possessions to appease the rain god. Beautiful things of gold and silver. And there were skeletons and skulls too from the human sacrifices. Human beings were the most valuable sacrifices of all to them."

He sat down beside her on the stone border. She was sitting well back from the edge. When they first came up she had walked out far enough to lean out and look down. There was nothing down there but a sheer drop to the rocky beach. She had always been a little afraid of heights. She tried to smile at him.

"It was dangerous," he went on. "It was a natural limestone well and it didn't go straight down but curved in places. It was so deep one of the crew got the bends even though he followed all the rules. They didn't have a recompression cylinder there so he had to be flown out in a U. S. Navy plane to a hospital that had one. And because the cabin wasn't pressurized, they had to fly low over the water all the way to Florida to keep from making him even sicker by going too high up." His brown eyes shone with excitement.

"Do you enjoy dangerous situations, Martin?"

He laughed merrily, as though she had paid him a compliment that he relished. "Well, there is no point in living one's life like a plant, is there? One has to take chances in order to be a man."

Ah, the old Latin machismo, she thought wearily. Manliness in the extreme. Or *in extremis*. It had always puzzled her how the Latin male could put their women on pedestals and yet regard them as being beneath them at the same time.

On an impulse she asked, "What do you think about women's lib, Martin?"

He made a face. "Don't tell me you are that type too. I can't believe it."

"Too?"

"Like Karen. That's her problem, you know. Pretty, but too masculine. She needs a good man, then she would forget all that nonsense."

Martin, it seemed, had a simple solution for everything. "How did you happen to come on this expedition, Martin? I mean, how did you hear about it?"

"I read of it in the newspapers at home. In Mexico City. It seemed interesting to me because I had never been to Europe before. So I sent an application to Professor Windle and after a long time I was finally accepted."

"Did you know any of the others before you came?"

"No, they all applied the same way and we met when we got here."

"How did Professor Windle decide who to choose, I wonder?"

"I don't think he had much to say about it, in fact. It was decided by the foundation who to accept or not. When we first got here, Professor Windle didn't even know much about us. He said our applications with our records hadn't been sent on to him yet, so we all had to tell him about our backgrounds."

"That's kind of odd, isn't it?"

He lifted his shoulders and let them fall. "I don't think so. Not if you know Professor Windle. He isn't interested in details like that. He only cares about the archaeology part. As long as we did our jobs, he didn't care who we were."

"This foundation is new, isn't it?"

"I believe so."

"Do you know who runs it?"

"Some very rich man, I suppose. He must be Spanish-speaking, because he insisted that most of us come from Spanish-speaking countries."

She remembered that Judy had mentioned that too. And she remembered all at once what Alex had told her. Krueger reportedly escaped to Argentina. Whoever he had sent back would almost certainly speak Spanish. Very convenient for him, the way Omega chose its student divers.

But she tried not to think about Krueger. If she got frightened, she might not be able to think clearly. "Martin, tell me, what do you remember about that day my sister was killed?"

He didn't seem surprised by the question. "That was what you were doing on the boat today, wasn't it? Trying to see exactly how it happened?"

"Yes."

"You must have loved your sister very much," he said, gazing at her thoughtfully. "Do you have other sisters and brothers?"

"No."

"And your parents?"

"We were both 'late' children. My father is retired now and he and my mother live in Florida." They were very content there. Her father liked to garden and play golf, and her mother had her bridge club and her charity work.

He nodded as though he now understood everything. "I have four brothers and three sisters," he said proudly. "All of them are younger than me. We live near the university in *la Ciudad de México.* My father teaches there. Have you ever been to Mexico City?"

She said she hadn't. When she and Judy were children their parents took them on a vacation to Mexico, but they only went as far south as Monterrey. After the trip Judy had decided to take Spanish in high school.

She hadn't forgotten that Martin still hadn't answered her question. "That day—" she prompted him. "I'm trying to visualize how it was. It's important to me."

"I understand, believe me," he said sympathetically. He leaned back against one side of the stone wall and crossed his arms over his chest. "It's something you feel you owe to her memory, isn't it? I respect you for that. I wish that I could have saved her. I deeply regret that I wasn't able to. Possibly if I had been standing nearer the winch at the time, I could have done it."

Even that wouldn't have helped her, Ann thought. "Karen was on the same side of the boat as you were, wasn't she?"

He cocked his head at the sky. "Let me see. She was lying down back there on the deck earlier that day, I remember that." He grinned slyly. "She had on her bikini, so naturally I noticed. That was earlier, though. Whether she was still there when I ran back I can't say. But I'm sure she must have been too far away to have been of help to your sister, if that is what you mean."

So Martin hadn't actually seen Karen at the crucial time, and Karen hadn't seen him.

"And the others, George and Eddie?" She added cautiously, "I suppose they were too far away to do anything either."

"That's right. I remember Eddie was filling a tank at the compressor. And I saw George jump into the water to take a swim a few minutes before. Just as though he hadn't gotten enough of the water already that day!"

That was something new at least, she told herself. George was already in the water when the winch was released. "He must have been on the wrong side of the boat too," she observed, merely to try to keep him talking about it. "The same as you and Karen."

"Ah," he groaned. "I can see what you are thinking."

Could he? She tried not to change her expression. But he said, "That it was destiny. Fate. If only one of us had happened to be near the machine when it broke down."

"Perhaps," she said softly.

His face went solemn. "Everyone with Spanish blood in his veins takes a fatalistic view of life. A tragic view. You know, it's even in our language. The way we ask 'What's happening?' is '*¿Qué pasa?*' Really it means 'What's wrong?' We seem to think, as a people, that anything that happens, anything new that comes up, has to be something bad."

"No, I don't believe in blind fate, Martin. When something goes wrong, there must be a reason."

"Ah, you American women are so logical. And yet some things cannot be explained. A certain part of a particular machine goes wrong on a certain day, and there is a death. A very nice young girl is killed. It's not reasonable, yet it happened."

She didn't say anything.

"It is difficult for you to accept, isn't it?"

"Yes."

He reached over and touched her arm, and his voice was as soft as rose petals. "But you know you will have to, sooner or later."

She stood up abruptly. "We'd better go, Martin. They may be waiting dinner."

NINE

Sitting at the long table with the others, Ann was gradually able to see Jorge and Eduardo emerge as something other than Tweedledum and Tweedledee. The meal was served in what the baroness called *die Grosse Halle*, the great hall. The name wasn't an exaggeration. The table they were sitting at was broad and long enough to serve fourteen in a pinch, yet it was dwarfed by the cathedral ceiling and the huge fireplace that looked like a small room. There was a grouping of heavy furniture near the opposite wall and a lot of bare floor in between. In the old days probably a hundred people or more must have lived at Schloss Adler, and this would be where they ate, with bare wooden tables lined up one after another.

A dramatic staircase cutting a diagonal across the wall opposite the fireplace led up to several bedrooms, one of them where Ann was staying. Olga von Toblen had shown it to her earlier when Ann brought in her suitcase from the car. It was a stuffy old-fashioned room with a canopied bed, dark polished wood floors, and a window that stuck. She could see the tops of the houses in the village from there. And down the long second-floor corridor was a bath even more antiquated than the Bruners'. Of course that made sense once she thought about it—the von Toblens would have been the first people in Tellin to put in the modern conveniences. No one but Ann was staying on the second floor.

Now Baroness Olga was presiding at the foot of the long table, making small talk in her heavily accented English. Her eyes darted from one plate to another to be sure each of her guests had ample portions.

Wilfred von Toblen was nearer Ann at the head of the table, and he was a contrast of subdued forbearance. He said very little and avoided eye contact with everyone, yet Ann got the impression that he was listening intently to what was going on around him.

Professor Windle, Karen, and Martin were there too. Karen had tied her hair back with a piece of twine and repeatedly made such rave comments about the bland, starchy food that Ann took it to be sarcasm. Every time she glanced at the professor he was looking at her with a mournful expression. She wondered if it had occurred to him yet that Judy's death might not have been a stupid error of some kind, but a planned murder.

Martin, at her right elbow, was giving her all his attention. He was almost a caricature of courtliness the way he had held her chair for her while she sat down and then passed each dish to her before taking anything for himself. "This is very tasty," he said once. "Try this."

But it was Eddie and George that she watched most closely, though as inconspicuously as possible, over forkfuls of veal and dumplings. Their personalities emerged. George was as withdrawn as the baron, but in a quite different way. Unlike Baron von Toblen he looked sullen, one corner of his mouth pulled up crookedly. He kept his head bowed over his plate, but she saw his dark eyes flick up to glance from one to another of the others from time to time.

But the boy she had thought of as Tweedledee wasn't reacting in the same way. Eddie chatted jovially across the table with Martin, speaking rapid-fire Spanish. At one point he even said something in Spanish to Karen as well, and Karen laughed and shrugged as though she couldn't understand him.

After dinner the divers went across the courtyard to their rooms. They were "in training," Karen said, and had to turn in early. The baroness invited Ann to play a game of backgammon, but she declined. It had been a long hard day.

She didn't get down in time to have breakfast with the others the following morning. They had gotten started earlier than she expected, and she was disappointed she had missed them. After having her breakfast she walked down the steep path to the beach. The dinghy was gone; she could see it tied up to the side of the boat. Figures in black wet suits moved on the deck. She couldn't tell the three men apart at that distance. Only Karen stood out because of her long hair. As Ann watched, Karen threw back her arms in an elaborate yawn and patted her mouth.

Ann decided she might as well sit down on the rocks and wait until the divers went into the water. She simply couldn't think of anything else to do.

She hadn't slept well. She had awakened several times and then she lay there in the strange bed thinking that she might never be able to find out who was responsible for Judy's death. She didn't even know how to proceed. She could only think of staying on for a while longer and talking to each of the divers, this time including George and Eddie, if she could get Martin to serve as translator. She wished now that she had learned at least the rudiments of Spanish as Judy had.

She was thinking that perhaps if she talked with each of them in turn, one of them might let something slip. But she really didn't have much hope of that. The person she was after would have to be cold-blooded, to murder a girl with four potential witnesses only yards away and make it look like an accident. If it hadn't been for what Windle told her, no one would have ever guessed.

Maybe it would have been better not to know the truth than to go through the rest of her life with the knowledge that her sister's murderer hadn't been caught. *No,* she told herself, *I've found out this much and I can't give up now.*

Still she knew that the longer she stayed at the castle the more obvious her suspicions would be. So far, she believed, Martin and the others had probably interpreted her visit as a simple desire to see the place where her sister had died and to talk to the people who had been there, to try to get all the details straight in her mind. As far as they knew, she still believed it was an accident. But that grace period couldn't be extended for very long. The longer she stayed at Eagle's Castle, the more they would wonder why.

And she realized that if she came too close to the truth—if the killer felt in danger of being exposed—she might be in danger herself.

Maybe I don't even care. She shook the thought away before it could catch hold in her mind. That was foolish. She came because it was something she had to do, not as a symptom of the unhappiness she had felt a few weeks before, after the divorce.

Two of the young men on the boat—she thought it was Martin and Eddie—leaped flippers-first off the stern and disappeared into the lake. As they jumped she saw vividly where the term "frogmen" came from. With their knees bent above those clumsy flippers and their eyes covered with the big face masks, they did in fact look like giant frogs.

Karen had moved out of view behind the cabin, and George was talking with Professor Windle on the bow. Odd, she thought, how she could tell now even at that distance that it was George. His head was hanging at a sulky angle, his body half-turned away from Windle as though he felt himself under attack and was about to run away. Yes, it was George all right.

She watched them for a moment, then scanned the rest of the lake. There was a lone sailboat far out on the water. She wondered if it might be Alex. She hadn't thought to ask him what kind of boat he used when he was out gathering samples from the lake.

Nearer shore, to her left, was a skiff with a blunt upturned bow. A man was standing in it pulling up fish nets that had been set out with keg floats. A few minutes later the boat moved closer and she saw that it was Max Bruner. She waved at him and after a moment's hesitation he waved back. He pulled in another net with a few silvery fish in it and then rowed off toward the village. She thought she could make out the roof of his house. He seemed to head the boat in that direction before he moved out of sight behind the rocks.

She couldn't see anyone on the diving boat now.

Probably they had moved into the cabin or onto the other side. Maybe Karen and George had gone into the water when she wasn't watching. *What is the use of this?* she asked herself in exasperation. She stood up, dusted off the bottom of her skirt, and climbed back up the stone steps to the castle yard.

She had just passed the tower when she saw Olga von Toblen coming toward her from the main building with a purposeful gleam in her eye. "Yes, there you are. I thought I saw you go down to the beach. I had thought after breakfast I should haff asked you if you might like to see the rest of Schloss Adler this morning. I would hate to haff you leaf without seeing the remainder of the interior."

It occurred to Ann just then that the baroness still didn't know she was Judy's sister and thought of her as a tourist. Perhaps it was just as well.

Ann went along with her, mainly because she was too preoccupied to think up a good excuse to say no. Very soon it became obvious that she was on the standard tour that Baroness Olga conducted for all her guests. Ann was led from room to room to the accompaniment of a commentary that grew almost singsong in places.

The halls and rooms of the main building seemed interminable. At times Ann, who had never had a keen sense of direction, got the wild notion that if the other woman suddenly left her she might wander for hours without being able to find her way to the front door. But gradually the plan of the building became clear.

The great hall took up one large corner at the front next to the part of the yard where the road led out to the village. Behind the great hall were the servants' quarters, a long row of small vacant rooms. Once there had been a dozen servants, the baroness told her, but now there were just two women who came in every day from the village to cook and clean.

They went through the kitchen, which looked out on a small bare yard. Then Olga took her down into the wine cellar, which she said had once been a dungeon. It was a dark moldy place with a slate floor set directly over the ground. There were two tall wine racks with many empty slots. The bottles that were there were dusty.

They came back to the first floor and went across the foyer to the other side, where they went past the von Toblens' living quarters. So far Ann hadn't seen anything of real interest, but the baroness had saved the best room for last. It was a sitting room near the end of the hall.

She opened the door and stood aside for Ann to go in ahead of her. The plastered upper walls of the room were completely covered by colorful frescoes. They were delicately painted scenes with pinks and greens dominant. Opposite the door a group of knights set out on horseback from the castle gate while women in long gowns waved farewell. The leader of the hunting party carried a falcon on an extended, padded arm. In a mountain clearing over his head a deerlike animal could be seen; the baroness told her it was a chamois.

Behind her across the room was a ballroom

scene, the action frozen in the middle of a step. A young woman in a pink dress glanced coquettishly at her partner. Over the doors of the end walls wood-cutters worked in the forest, and peasant girls led gaily decorated cattle up to the mountain *almen,* just as they still did every spring. There they would stay during the summer while hay was gathered on the lower meadows, and in the fall the cows would be brought back down to the barns. And over the top of the door they had come in there was a painted scroll with the legend, *Falken der Tugend.* Falcons of Virtue. The motto of the von Toblen family, she realized.

She was impressed by the frescoes. "When were they done?"

"Wilfred's father commissioned them shortly before the First War." She was transparently pleased by Ann's reaction.

"Very lovely."

There was one more door at the end of the hall. "Now we go out. But be careful in your step, please. This is the chapel, one of the unrestored areas."

The room was large enough to serve as a counterweight architecturally to the great hall at the other end of the building. There was no furniture and the floor was dirty and uneven. A pattern of light on the floorboards made Ann look up. There was a large ragged hole in the ceiling on the back side of the building.

"The Nazis used our chapel as a barracks. Such beasts," the baroness said heatedly. "Then the

Americans came. One of their big guns did that." She pointed at the hole in the roof.

"You and your husband weren't here then, I hope?"

"No, in Vienna, thank Gott." Her expression hardened. "They had put us both in prison at first in 1939, but they let us out for some reason. It was a miracle, for they knew my mother was Jewish. And my husband had been speaking out against Hitler since 1932. It was a miracle they let either of us go.

"The castle was in such condition after we returned, you cannot imagine it. The American soldiers had just gone. They had left the courtyard dug up in some places. They even dug in the cellar, looking for gold they thought the Nazis had left behind. Such foolishness; they didn't even find it. They should haff had the courtesy to leaf everything as they found it."

They went out by the large side door and crossed the courtyard in silence. In the other building, the *Unterschloss*, there was another surprise. A hall of armor, which paralleled the arcade, contained a modest collection of antique weapons and armor. There were three suits of armor standing along one wall with visors down, and battle-axes, maces, and spears arranged on nearby shelves. A musty smell suggested the room hadn't been opened recently.

"Upstairs are the other young people's rooms of course. And the professor is staying in the tower over the big workroom. You haff seen the tower?"

"Yes."

Baroness Olga began to recite the history of some of the antique weapons, but Ann barely listened. *Why is she making such a big production out of this?* Ann wondered. They went back outside. "My husband and I are especially proud of the fresco room," the baroness was saying. "He and I hope eventually to be able to restore the entire castle, but such things take time, naturally. We only haff begun receiving guests this past year." Then she began to talk about how she hoped the roads to Tellin would soon be improved, how she had heard that American travelers had grown weary of frantic tours going from one European hotel room to another.

"They are looking for something a little different, perhaps. American travelers would be interested in seeing an old castle, don't you think?"

So that was it. Ann was amazed. The woman was planning to try to turn Schloss Adler into a full-fledged hotel, a tourist attraction. And she was probably right, tourists would very likely come to the castle in droves if they heard about it, if the advertising were handled right: "Spend an unhurried week in a sixteenth-century castle overlooking a secluded Tirolean village. . . ." She wondered what Baron von Toblen thought about all this. She couldn't imagine him becoming an outgoing host to a large group of strangers.

The woman was waiting for her to say something. The tour of the castle had apparently been a device to learn Ann's reaction to the place. "Yes, I'm sure many Americans would be interested in coming here." They made their way back to the main house.

She was grateful that it was almost time for Alex to come pick her up for their lunch date.

TEN

She noticed a picnic basket on the back seat.

"Have you ever seen an *alm*?" Alex asked her.

"Only in *Heidi.*" She had just had time after leaving Olga von Toblen to go up to her room and change into a fresh skirt and blouse, a dark green outfit this time, the first thing she saw. And he had dressed informally too in a lightweight navy sweater, gray slacks.

She was impressed by the way he negotiated the Mercedes around the hairpin curves on the way down. Instead of continuing on the road into the village he took the sharp turn back toward the mountains. Before long the castle and the village were behind them. The road twisted through evergreens and rose steadily. They passed hay sheds and two farmhouses set back from the road. The trees became fewer. Eventually there were only gnarled stone pines and stunted green-black fir. After twenty minutes they reached a broad sloping meadow.

He parked beside the road. There was a herd of dun cows grazing seventy or eighty yards away and about the same distance in the other direction, a log house with a gently sloping roof, wide overhanging eaves, and a broad front porch. Smoke was trickling out the chimney. Higher up there was the bare angular rock of a mountain with snow at the peak—glacier snow that never melted, and beyond that only sky.

When they got out she could hear the cowbells, each one a different tone. They reminded her of wind chimes. They walked through the grass, skirting around an outcropping of rock, until they had a splendid view of a green valley a mile or so below. Alex spread a blanket on the ground for her to sit on.

She leaned back, resting on her hands, and he stretched out beside her. "This is perfect," she told him. It was peaceful, beautiful. She wished she could stay there for a long time and put Eagle's Castle out of her mind.

He guessed that she wasn't ready yet to talk seriously. So he said, "I like to come here. My grandmother used to have an *alm* very much like this. I would spend my summers with her." He smiled, remembering something. "I think I was pretty spoiled by the time she got her hands on me, but she set me straight very quickly."

"How do you mean, Alex?" It was difficult for her to think of him as a spoiled child.

"Oh, in Vienna I was put in a private school, and I couldn't help noticing how my teachers were rather impressed by my parents' name. My grandmother wasn't so easily impressed. She let me know early I'd have to make my own way.

And she taught me to enjoy hard work, the feeling you get when you've accomplished something."

"She sounds like a very nice lady."

"She was one of the best."

Ann took a deep breath. The warm sun and fresh air were already making her feel less tense. "I really love this place. Who does it belong to?"

"Max Bruner."

"He's a good friend of yours, isn't he?"

"Yes, I met him the first summer I came here, several years ago."

She remembered something. "He told me to remind you of your chess match."

"Oh, that." He laughed. "We have a small bet going and he's a pawn up on me. I'll let him finish me off one of these days." He glanced at her. "Do you play chess, Ann?"

"A little. I'm not very good, though. Somehow I think it's more of a man's game." She leaned her face back against the sunlight and closed her eyes for a minute.

"You could be good at it, I'm sure."

"There's never been a really great woman chess player, has there?"

"Not that I know of."

"I wonder why." She was happy to get the chance to think about something else.

"The traditional answer is that women don't have analytical minds, but it's probably a matter of cultural conditioning."

"Maybe women don't want men to know that they can plot out every move well in advance, instead of acting on the spur of the moment. Women are supposed to react to what men do, aren't they, instead of thinking up intellectual game plans of their own."

He looked at her with a quiet interest. "Do you plan your moves well in advance?" She knew he was talking about her emotional life.

"Sometimes I wish I could." She didn't want to talk about that. "That's why I'm no good at chess."

He let the conversation revert back to safer ground. "You'd probably rather read or play tennis, I imagine."

She did like reading and tennis, as a matter of fact. "Yes, how did you know?"

"Oh, just a guess." After a moment he said, "Are you hungry?"

"A little." He opened the basket and pulled out a long loaf of bread, cheese, white wine. They ate for a while in silence.

"Alex," she said all at once, not looking up. "I found out yesterday that Judy's death couldn't have been an accident."

She told him about her experience on the boat, the experiment with the winch, while Alex listened intently. "The chamber must have been three or four times as deep on that afternoon as it was supposed to be. Someone deliberately lowered it to ensure that she wouldn't survive."

"I see. Ann, I'm sorry."

"I know. But at least I know now how it happened. Though sometimes I wish I didn't."

"How did you get the idea of bringing the chamber up?"

"When Windle told me it had taken three or four seconds, it didn't sound right to me. When I was at home trying to find out what had caused Judy's death, the actual physical reason, the people I asked kept emphasizing how the chamber must have shot up with terrifying speed, so quick that her body didn't have a chance to adapt to the change in air pressure. I don't think Windle ever thought about it because he was so certain it was at ten feet. Anyway, the time involved just happened to come up when he was talking about it."

"Still, it was bright of you to think of it. And yet not so bright," he added quietly, "if the divers all know now that you realize it wasn't accidental."

"No one knows, I'm pretty sure, except Windle. No one else was around when we were discussing that part of it."

"Even so." He looked concerned. "Do you have any idea who tampered with the winch?"

"No, everybody claims to have been in some other part of the boat at the time."

"That was one reason, I think, that the police were convinced there was no foul play involved."

"But it *must* have been deliberate; you can see that." Her voice rose a little.

"Of course, I agree," he said quickly. "You're sure Windle is right about the three or four seconds it took?"

"I'm certain he is. If the chamber had come up immediately the way it did yesterday, it would have been all over before Martin could have run across the deck to jam that lever back down. But it wasn't."

"Martin Cabrera?" She had forgotten to mention that part of it.

"Yes, he was just heading that way, and when he heard the cable reeling out he ran across and tried to stop it."

"And he didn't see anyone around the winch then, I take it?"

"No. I've been wondering about that too. There was a lot of machinery over there to block his view. Someone might have slipped out of sight that way. Or dropped over the side of the boat into the water. It wasn't very far. George was already in the water when it happened, they say. There's also a passageway around the front of the boat that Karen might have used if she were the one. I don't know. None of them really has what you could call a good alibi."

"And where was the other diver?"

"On the stern, apparently. Martin says he saw him back there. But he didn't say *exactly* when. And I'm not certain about Martin's story either. No one actually saw him running toward that winch."

"Martin Cabrera. Yes, I remember him now. I'm afraid I've never liked that name much. It used to be popular in this part of the world because of Martin Luther. But now it carries the memory of Martin Bormann for too many people."

"That's funny, I never associated his name with

Bormann at all." She said, "In America the Second World War is like ancient history to most people, I guess. The past is so much closer here. Here it seems—always a step away."

"That's very true." And then, "Dammit, I was afraid of something like this even before I read those letters your sister sent you. I kept hoping I was wrong."

"I know. Whoever it was, it must have taken more time to locate those boxes of gold than he had thought, and so he had to keep going off alone. I suppose if they're in the lake they must be completely cøvered over with silt by now." She was talking slowly, thinking about it.

"So you do believe it now, that Krueger has sent someone?"

"Yes." Nothing else made sense to her. "One of the three Latin-American boys, or Karen, must be working for Krueger."

She took a small sip of the wine and set the glass down carefully on the blanket. She said, "I found out something else that you might want to know about. Or possibly you already know. The applications the divers sent in were all handled by the Omega Foundation. Professor Windle didn't have anything to do with choosing them, apparently. I was wondering if Krueger could be mixed up with Omega. If he did have something to do with it, he might have been able to make sure the person he wanted to send to Tellin got picked to work with the expedition, don't you think?"

"You're right. The police have been wondering

about Omega too. It turns out that it's a paper organization. The address was a box number in Geneva, and the address given was a vacant office building. They've tried tracing it through bank records, but that has led only as far as an unnumbered account. That's something I heard about from Kurt yesterday."

The famous Swiss bank account. Secret funds, untraceable. "He must have made a whole new life for himself in South America. He could have accumulated some money for just this purpose."

"That's the idea."

"Alex, how many people are working on this with you?"

"Several people from the department in Vienna. They've set up a temporary headquarters at Innsbruck. And there are the investigators overseas looking into the backgrounds of the people involved."

"Have they found out anything new?"

"Nothing of substance since I talked to you before. Kurt claims they've tracked down one false lead after another."

"What about here in Tellin?"

"Besides me, you mean?" He finished off a piece of the cheese and wiped his hands on a napkin. "They were afraid to send any undercover policemen here. In a place this small any stranger would be conspicuous. Max has been helping me. He would like to get Krueger for personal reasons. He worked with an anti-Nazi underground group here in Tellin during the war. He's out on his boat today keeping an eye on the divers from that angle."

She recalled the scene in the Bruner kitchen. "His wife must know about it too. She looked upset when she heard I was going to Schloss Adler."

"I wouldn't be surprised that he told her."

She was glad she had Alex to talk to. If she hadn't happened to meet him on her way into Tellin she wouldn't have known about any of this. She would have gone straight to the castle without realizing what was going on. She was thankful that he had been willing to trust her, and not let her walk in there without knowing.

"I saw Max on the lake this morning. I thought he was simply out fishing."

"He probably was doing that too. He has fished those waters for years. All the fishermen in the village have their own particular section of the lake staked out, and the waters near Schloss Adler happen to be his. I'm sure he'll be glad when this business is all over."

She had been watching his face, and something about the way he said that made her remark, "You'll be glad too, won't you? I don't think you like this assignment very much."

"I wish I weren't involved. I feel partially to blame for your sister's death. They assured me there would be no danger to anyone at Schloss Adler if we simply waited for Krueger's agent to find the gold and bring it out. But it didn't work out that way."

"But it wasn't your fault, Alex. It was no one's fault but whoever did it. And Heinrich Krueger's.

What else could they do but wait?" But she could tell she hadn't convinced him.

"Besides, some of the methods they're using bother me. I don't like subterfuge, and I've let myself be pressured into doing things I don't like."

She was about to say something when she saw him glance up sharply toward the distant house. When she looked around she saw a middle-aged woman coming out the back door carrying a basket of clothes. She was wearing an old-fashioned dirndl, the first one Ann had seen since she arrived in Austria. She went to the line behind the house and began hanging out the wet clothes. She moved woodenly, like someone much older.

Ann turned back to him and before she could ask who it was she saw that the serious look on his face had deepened into something else. "That's Freida, Max's daughter," he said quietly. "She stays up here all summer and fall, right up until the heavy snows. I've never seen her come out of the house before."

Ann looked back at the strange woman. "Shouldn't we go over and speak to her before we leave?"

"It would only frighten her. She wouldn't have come out if she had seen us over here."

"Why not?"

He pulled a handful of grass and threw it down. He seemed to be undecided whether to answer her. "You said the past is closer here and it's true. Especially in the Second World War. By the time the German soldiers got to Tellin they knew the war was over. Several of them got drunk one day. They caught Freida working alone in the meadow below the castle. They beat and raped her. She might have gotten over it eventually, except that she had been born retarded, and so it was devastating for her."

After what seemed like a long time Ann said, "It must have been her photographs that I saw when I was staying at the Bruners'. They went through her childhood until she was in her teens and then they stopped." She found herself barely speaking above a whisper even though the woman was much too far away to hear them.

"That's when her life stopped, in effect. Max and Elsa must have sensed that. No more pictures. She wouldn't have wanted it, anyway. Afterward she never wanted to go out of their house. She couldn't face the people in the village. Max sent her away to live with relatives in another province for a while, but that didn't work out well, either."

"Were those soldiers ever caught? Punished?"

"Naturally Krueger didn't do anything about it. He regarded all Tirolese as inferiors. Especially someone like Freida. The Nazis called the retarded *Unnütz Esser*, useless eaters. Many of them were exterminated like the Jews. By the time the American troops got here everything was in confusion. Many of the Germans had taken off their uniforms and slipped through the lines. Max tried to locate those men after the war, but he finally gave it up as hopeless. I think it still haunts him that they got away. If he could have found them, he would have killed them, I'm sure. And that's why he volunteered to help the police catch Heinrich Krueger."

She had heard the whole litany of Nazi war crimes but they had never seemed real until now. Now she saw the truth in human form, in a middle-aged woman hanging out clothes. She heard him say, "This part of Austria has been a crossroads for armies for over two thousand years. The Roman legions, the Lombards, and the Franks. There must have been a lot of Freidas during all that time. History keeps going around in the same damn circles."

Then he said, "And so the past is only a step away here, as you said. We can see the past around us in broken lives."

"That's why you agreed to help the police, isn't it? Because of people like Freida."

"I never actually thought it out in those terms, no. I simply think it would be a terrible injustice if Krueger were to get away with that gold."

So do I, Ann thought. The woman finished hanging the last towel and went slowly back toward the house. She hadn't glanced their way. "She must be very lonely. I wish I could have spoken to her."

"She wouldn't have wanted you to, believe me. I'm sorry, Ann. I've made you sad, telling you that story. I didn't expect that we would see Freida today."

It was true that she was sad. The light falling across the *alm* seemed different, even the jangling of the cowbells had fallen into a minor key. But she was glad that he had told her about Freida.

On the drive back he tried to persuade her to let him take her to the Bruner house again. "You can still go up to the castle tomorrow, if you insist. I'll go with you. We could talk to the divers together, if that's what you want to do."

"Even though it might jeopardize what you're trying to do?"

He sighed. "Yes, if that's what it takes to get you out of there."

She didn't have to think about it very long. "I don't believe it would be a good idea, Alex. We're after the same person, I'm sure of that now, and I don't want to scare him off any more than you do. As long as I'm staying there it would seem natural for me to talk to them, but to make a special trip up there to ask them questions? No, I don't think so."

"Then let the police handle it. When they uncover whoever it is, he'll be charged with your sister's murder along with the rest of it."

"But suppose their investigation doesn't turn anything up?"

"Even if it doesn't, every one of them will be stopped at the border, searched and questioned."

That was something, at least. "I've been wondering how anyone could hope to manage to take all that gold out of the country. Alex. What did you say it was—around five hundred pounds?"

"Yes, somewhere around a quarter of a ton, they think."

"Where on earth do they expect to hide it once it's brought out of the lake, if it's as much as that?" She imagined a huge pile of gold stacked up on the shore beside the castle.

He gave her a quick grin. "You don't edit any chemistry texts, I can see that. Pure gold, remember, is very dense, almost like lead. A single bar around six inches long, which would be worth about $30,000 now, would weigh almost fifteen pounds or so. You can put a quarter ton of those bricks in a very small space, something like a foot deep and wide and one and a half feet high." She saw the huge pile of gold covering the beach shrink down to the size of a footstool.

"It would get lost in the trunk of a car," he continued, "if it weren't for the weight. Of course, I'm talking about pure gold now, and this may not be. In fact, it probably isn't. Some of the other Nazi caches that turned up were partly alloyed gold objects that had been hammered flat. Things like that would be bulkier and lighter. They aren't sure exactly what this one will turn out to be, but space would definitely not be a problem." It was obviously old ground he was covering with her. He had probably already discussed all this thoroughly with Kurt and with Max.

"Then it could be hidden almost anywhere," she said. "If the pieces were small enough, they could be brought out of the lake a few at a time inside one of the divers' wet suits and hidden—"

Where? she wondered. "They'll start crating up their equipment before long, I suppose, getting ready to go back. It could be stashed inside some of those crates."

"Yes, but remember that Windle will probably be around to supervise the packing. Then he'll have the trucks he rented before to bring the stuff here come back from Innsbruck to load up and take it to the airport."

"And it would probably show up at the airport, I guess, when they weighed everything."

"Right. Or the inspectors might decide to open them up."

"So you don't think they'll do it that way?"

"My guess? Too risky. Besides, the divers aren't supposed to fly back with Windle and the gear and what few artifacts the Austrian government will let him take out of the country with him. They're supposed to return to their own countries, so none of them would be around when the crates are opened. They would probably have to divert the right crates before they were flown to the United States, somehow."

"Through someone working with the airlines, maybe?"

"There's an outside chance of that, I think, but there's an easier way to do it."

She was trying to guess what it was. "You mean not bring the gold up to the castle at all? Just bring it to shore and hide it somewhere?"

"Not even bring it out of the lake. Simply move it from wherever it is to a place very near the shore and farther away from the village. Then when the time came, I imagine a vehicle of some kind would be driven to some place nearby and they could simply wade out and haul it in. At night, probably."

"And then hide the gold inside the vehicle. Yes, I see. It shouldn't be too difficult to cross the border with it concealed inside a car or a truck."

"Ordinarily the inspection at the frontier wouldn't be rigorous. It wouldn't have been difficult at all to get through if it hadn't been for that anonymous tip. Whenever I'm out on the lake I search the shore line for a fairly level spot, a place without too many trees, where a vehicle could be brought down fairly close to the water. So far I've found a couple of spots, and the police have assigned men to watch those areas as inconspicuously as possible. And even if there should be a slip-up there, once the divers leave Tellin they'll all be followed and stopped at the border. Every vehicle going out of here will be stopped."

He was trying to tell her she didn't have to worry, her sister's killer wouldn't get away. "I understand what you're saying, Alex. But something might still go wrong."

"Ann, I haven't told you all of this in order to involve you further. I've told you because I want you out of it."

She could look at him and tell that he was very worried about her. "Alex, I know I'm making everything more difficult for you. I'm getting in your way, I guess. But I'd feel the way Max Bruner must feel; it would always haunt me, if I hadn't done whatever I could. I want to stay there just another day or two, at least. I might be able to find out something that you couldn't, just by my being there." She paused. "Even if I don't, at least I'll have tried."

He wished there were some way to reach her. "Is it going to help your sister to risk putting your own life in danger?"

"I'll be careful, Alex. I don't feel unsafe at Schloss Adler, really. When I do, I'll leave."

He took his eyes off the winding road long enough to give her a searching look that showed exasperation but more than a little admiration as well. "You are a difficult woman," he said.

And then, trying to lighten the mood for both of them a little, "Maybe I should have said 'determined and strong-minded.'"

She smiled because she remembered that. "Stubborn is the word, I guess."

ELEVEN

The next morning she went out to the boat with the divers, saying something as they got into the dinghy about wanting to see how an underwater dig was carried on before she left. She stood around while they got ready for the morning dive. All four of them pulled on their tanks, adjusted their face masks, and jumped overboard feet-first, one after another. Then there was nothing to do but wait until they came up again.

She had coffee with Professor Windle in the cabin, and he told her how he had been a supply sergeant during the Second World War. Not a very good one, he said. He kept forgetting where everything was. Then they chatted about the latest news from the States that had just come in over the radio. Another fluctuation of the dollar had sent the price of gold upward again.

Another morning wasted, she felt.

They all had lunch on the boat, roast beef sandwiches and coffee. Martin and Karen talked about how the work was going and about the weather. The crew had long since finished with the first level and were almost through the second stratum, the last one they planned to do that summer. All the photographs had been taken; the charts showing the exact location of every item had been drawn. They were bringing up the last few artifacts that week. Then they would take down the grids, finish the cataloguing, and start to pack up.

During a lull in the conversation Ann asked, "What is it like down there? Is it cold?"

"Damn chilly," Karen said. She looked more glum than usual that day, Ann thought.

"Is the site directly under the boat or off to one side?"

"A few yards over that way," Karen told her, gesturing lazily toward the side away from the castle.

Ann had been looking at the shore line above

the castle that morning. Beyond the tall rocks around the beach, trees came directly down to the edge of the bank. There were cliffs in some places, and what looked like quiet coves in a couple of others. She couldn't tell at that distance whether a vehicle could approach those places or not.

"Have you ever been scuba diving, Ann?" Martin asked. He was being as attentive as ever, watching her with his soft brown eyes.

"Once on a vacation in Bermuda." She had been fascinated by the way the colors changed as she went deeper. First the reds disappeared, then the oranges and the yellows. Everything was blue and green for a while, and finally there was nothing but a cool clean blue.

"Then you know the fundamentals," Martin told her. "Maybe you could come down with us this afternoon and we could show you the site."

Sitting beside her on the deck, Professor Windle tensed up perceptibly. "That may not be a good idea, Martin. It might not be safe."

"Thanks anyway, Martin," she added. "I really don't think I'd care to." She preferred to look at the site the way Windle did, through photographs. Besides, she wasn't sure she trusted Martin, in particular.

That afternoon she spent part of her time looking over some of the shots Windle had spread out on the desk in the cabin. There were more pictures of the brown lake bed with metal grids and numbers, very much like the ones she had seen in the tower. A few close-ups of special details. They didn't tell her anything she wanted to know.

"Professor Windle, I've been curious about the Omega Foundation. Could you tell me something about it?"

"It's a new outfit, and I'm afraid I don't know a great deal about it myself. They approached me in a letter last winter." Looking rather absent-minded, he opened the long drawer of the desk and rummaged through some papers. "Yes, here it is." He pulled out a sheet of paper with some typing on it and glanced at it before handing it to her.

It was written on good bond paper. The printed letterhead identified the Omega Foundation and gave an address and box number in Geneva. It read:

My dear Professor Windle,

We have learned of your interest in conducting an archaeological expedition at the newly discovered Bronze Age site under the lake at Tellin, the Tirol, Austria.

As a newly formed foundation interested in promoting the study of underwater archaeology, we would be very pleased to discuss with you the possibility of our providing the financial backing for this expedition. Since we are primarily an educational foundation, our only stipulation would be that we be allowed to play a major role in the selection of students who would assist you in this endeavor. We are particularly interested in encouraging the training of underwater archaeologists from the Latin-American countries, but other qualified applicants would also be considered.

If this proposal meets with your approval, please contact me at the above postal address at your earliest possible convenience.

It was signed William R. Samuels, Director of Project Planning.

"Did you ever speak to Mr. Samuels?" she asked.

"No, we exchanged a few more letters." He looked a trifle embarrassed. "I did have the foundation checked out of course. They *are* very new, but they did seem to have a good deal of money deposited in a Geneva bank, so I assumed they were reliable. They've financed us quite generously, I must say."

"If it's not too personal, Professor Windle, how much does a project like this cost?"

"This one will run, oh, somewhere around a quarter of a million dollars."

She tried not to show how surprised she was. Obviously someone had wanted very badly for the Tellin expedition to get off the ground.

Shortly before dinner there was a knock on her door. It was Karen. She came in, went directly to the bed. Flopped down, leaning back on her palms. "So how are you getting along?"

"Okay." Ann felt uncomfortable standing over her. She pulled up a chair.

"I wanted to talk to you before. About Judy."

Ann waited.

"She was a nice girl. I liked her. We roomed together, as you may know."

"Yes, she told me. Wrote me, actually." *What did she want?*

Karen eyed her for a moment, then said, "So you've done some diving yourself?"

"Just once, like I said. Judy and I took lessons together at a hotel in Bermuda."

"That's how your sister got interested in it, I suppose."

"She loved it right away, more than I did." There was a pause. "How did you get started diving, Karen?"

"A boy friend when I was in high school taught me. I was crazy about it from the beginning too. I love the feeling of freedom you get, being completely cut off from the real world for a while."

"Is this your first archaeology expedition?"

"Yeah. I'm not even an archaeology student like the others. Strictly an amateur."

"Then how did you happen to get involved?"

"Oh, I thought it might be fun, after the other one."

"The other one?"

"The only other dive I've been on. It was out in the gulf near Tampa. A group of guys from Michigan had formed a corporation to explore a sunken galleon they thought contained pirate treasure. Just our luck, it turned out to be a dud. Nothing of value at all except a few historical things. I left before they finished, anyway. I couldn't stand the atmosphere."

"Why not?"

"Oh, it got to the point where everybody was acting paranoid. The leader was uptight because his divers kept coming up empty-handed. He started acting like he thought we were stashing fistfuls of doubloons in our wet suits and ripping him off. But of course there was nothing to rip off. Anyway, I decided to get out. I thought *this* dive would be much tamer, just a nice relaxed way to spend a summer. Besides, well, I got into a bad scene with the guy I was living with last winter." She let that drop.

"It must be nice to do something you really love to do and get paid for it besides."

"Yeah, that was a big surprise, the salary. Nice."

Ann was feeling more confident now, ready to steer the conversation to the main point if she could. "My sister was enthusiastic about this project at first. But for some reason I got the idea from her letters that she wasn't very happy toward the end."

"Is that right?" Karen pursed her lips, thinking it over. "Could be. I noticed she was moody sometimes."

"Was there anything specific that she said?"

Karen sat up, folding her hands around a knee. "Not that I remember." Then she said, "I did notice, though, that there were some bad vibes going when we were all together. Between Judy and one of the Latinos. You know how it is when some people just rub each other the wrong way."

"Which one of them was it?" She tried not to sound eager.

"Eddie, I think."

"She never said anything about it?"

"Nope. It was just a feeling I had. I have very good ESP."

"Eddie is the one from Chile, isn't he?"

She hesitated a moment. "Well, that's what he *says.*"

Ann felt her heartbeat jump. "Why do you say that?"

"Oh, just something Martin said to me once. He said Eddie's language didn't sound right to him, that he sounded more like he was from Argentina."

As casually as she could manage, she said, "From Argentina? I don't understand what you mean. Spanish is Spanish, isn't it?" She hoped she hadn't spoken the words too quickly, hoped her voice was calmer than she felt. It was the first break she had had.

And it was the first time she had seen Karen smile. "Oh, not at all. Every one of those countries has its own small differences in the meanings of words, just like there are differences between U.S. English and British English. In Mexico a word might mean one thing, and in Panama something else. But I've forgotten the example Martin gave me now."

"You understand a little Spanish?"

"Yeah. Being from Florida, you know, I picked it up from some Cuban friends. They have their own special dialect too, they say."

Ann was very aware that she shouldn't give away too much. Maybe she had already sounded too interested. "Well, I wouldn't know about that," she said.

"It's not important, anyway." Karen looked unconcerned, almost bored. "Martin may have been jiving me, for all I know. But he may have told your sister the same thing about Eddie. That would be like him, flitting from flower to flower with the same line of pollen."

"Why would he do that? Make up something like that, I mean."

"I wouldn't know. Maybe he saw Eddie as a rival, and he wanted to cut him down. I think Martin had his eye on Judy. Strictly as a sex object, of course. I only mentioned it because I thought it might have something to do with why your sister and Eddie didn't hit it off. She wouldn't have liked a phony any more than I would." After a pause she added, "You two were pretty close, I can see that. You know, when you first came here, I thought I understood. Your coming to see exactly what the accident was all about. Then I got to thinking about what happened on the boat. Windle bringing the chamber up, I mean. And since you're still here today, well—"

Ann was careful to keep her face a blank.

"You must think there was something kind of off base about the way she died."

"Do *you* think there was?" Ann asked quietly.

She studied Karen's face carefully. Good bones, clear skin. Brown eyes that never let you get too far into them. She was about twenty-four. Born about 1953, she would have been around ten years old when John Kennedy was killed, fifteen when Martin Luther King died. . . .

"I honestly don't think so. But nothing would surprise me. It could happen. Anything could happen." She got up and went to the door, without any apparent feelings about it one way or the other. "Listen," she said, "how long will you be staying?"

"I don't know. A day or two, anyway. I really am interested in finding out more about the work Judy was doing here."

"Well, if there's anything I can do to help you out."

"Sure."

Karen started out the door, then turned back. "Maybe I haven't shown it, but I really *am* sorry about Judy."

The male divers didn't show up for dinner. Karen moved her plate down to sit by Ann. The three young men had taken the van into the village. Since there was no diving scheduled for the following morning, Windle explained, they had decided to have a "night on the town." They would probably be late getting back.

"Typical men," Karen said briskly. "They didn't even think to ask if *we* might like to get out of this place for a while." Then, under her breath to Ann: "Frankly, much more of this damn food and I think I'll turn into a potato." Still, Ann couldn't help noticing that Karen ate heartily. She remembered how her sister told her that diving always gave her an appetite.

After the meal Professor Windle excused himself to do some cataloguing in the tower workroom, and Karen said she was going across to her room to read for a while before she went to bed. Not wanting to spend the evening trying to think of something to say to the baron and his wife, Ann went upstairs to her room.

She had been trying to decide what to do next. Tomorrow she might have a chance to ask Martin about Eddie. If he confirmed what Karen had told her, what then? She couldn't talk to Eddie herself without using Martin as an interpreter. And she wondered what good it would do. She didn't know enough about either Argentina or Chile to be able to ask questions that would reveal anything. What if he really was from Argentina? What if he was the one they had been looking for? She might accomplish nothing with her questions except to put him on guard.

She decided to go see Alex the next day and tell him what she had found out. Maybe if the police investigators concentrated their efforts on Eddie they would be able to find a flaw in his story about being from Chile and prove something one way or the other. That was probably the best way to handle it.

But in the morning there was something new to worry about.

TWELVE

At eight-thirty she came down to the great hall and saw that the only people having breakfast were Baron Wilfred and his wife. They looked rather ludicrous sitting at opposite ends of the long table as though they were entertaining an invisible party of ten or twelve. Baron von Toblen was concentrating on his plate, looking dignified, solemn, lost in thought. She wondered if the others were sleeping late since they didn't have to go to the boat.

The baroness said good morning, and when Ann asked where the others were she shrugged and waved vaguely toward the front door. Instead of sitting down, Ann went outside.

The four of them were standing near the tower. As she started over that way she saw that the three young men were listening intently to Professor Windle. Before long she could see that Windle's face was full of worry and bewilderment. When she came up to the tight circle they had formed, George and Martin stepped back without a word to include her, and this easy deference alone told her something was very wrong—it reminded her too much of the elaborate concern for others that people always seem to show when a disaster strikes. Neither one had actually looked at her, they had merely stepped back to let her join them, and they all looked stunned.

And then Professor Windle noticed her and said simply, "Karen has disappeared."

It was then that she realized that the van wasn't in the courtyard. Windle and Martin took turns telling her the rest of it—as much to lay all the facts out for themselves, apparently, as to inform her.

Martin spoke first. "It happened late last night. The other boys and I got back from the village about midnight and went up to our rooms. Karen must have been asleep, because I noticed there was no light under her door. Then after I went to bed, about a half hour after that, I heard someone rapping at her door and calling her in a soft voice. 'Fräulein Phillips?' he said. And I wondered, 'Who the devil is that?' I was about to get up and go ask him what he was doing there when I heard her open her door and say, 'What is it?' She sounded like she just woke up. And I heard him say, 'An emergency telephone call has come for you in the village, fräulein. You are to call back at once.' And then she said something I couldn't hear and he answered her, and he said, 'It came just a few minutes ago. That's all I know about it. I was to tell you to call back immediately.' And then a moment later he told her that he had walked up from the village to deliver the message to her, and she said, 'Wait a minute.' "

Martin was speaking quickly, glancing occasionally at Windle, as though for confirmation. "She came out into the hall," he said, "and walked down to my room and opened the door. She came over to my bed and was about to waken me, I suppose, when she saw that I was already awake. I sat up in bed, and she said, 'Give me the keys to the van. I have to go.' So I showed her where they were on the dresser and she went out. She and the man must have driven away in the van. I heard it start up and go off."

George said something in Spanish. Martin said, "George and Eddie say they didn't hear any of this last night. That makes sense, because my room is between theirs and Karen's."

"But the odd thing is, there *wasn't* any phone call," the professor said excitedly. She noticed now that Windle was breathing rather quickly and that his cheeks were blotched with pink as though he had overextended himself.

Martin spoke up again. "I went back to sleep after that, but I woke up very early this morning, still wondering what had happened. When I went to Karen's room and saw she hadn't come back, I went to the tower and woke up Professor Windle and told him. When she still hadn't returned by seven o'clock, he walked down to the village to try to find her."

"And she wasn't there," Windle said. His eyes looked larger than ever behind his thick glasses. "She didn't leave a note or message, or anything. No one down there saw her last night. I looked everywhere. I just got back when you came over here, Ann."

"Then it must have been a serious emergency back home," Ann suggested. "She may have driven on to Innsbruck to catch a plane to the States."

"That's just it, Ann. I checked the various places where the phone call might have come in to try to find the man who delivered the message. There are only two phones in Tellin, I soon found out, so it didn't take long. There's one at the inn and another at the post office that I already knew about. The post office was closed last night of course, but it seemed the logical place for an emergency message since there's no phone up here. And the postmaster could have heard the phone ringing downstairs from his living quarters overhead. But he said there was no such call. At the inn they said the same thing. No one knew anything about it." Windle looked stricken. "Karen has simply disappeared."

She said, "But maybe the person who took the call just didn't want to be involved any further. Possibly," she added, warming a little to her idea, "with all due respect, Professor Windle, maybe the people at the inn or the post office simply didn't think it was your business to know about a call Karen got."

"I wish I could believe that," he said dismally. "But I told them it was important that I find Karen, that I was very concerned about her because I didn't know where she was. And everyone I talked to seemed genuinely mystified by my question. In any case, I took the precaution of calling the nearest police station in Zirl. Not much help there either, I'm sorry to say, because they claimed they couldn't send someone out looking for her until she had been missing for twenty-four hours. But they did say they would check the airlines in Innsbruck and Salzburg to see if she made a flight early this morning. So there's nothing to do now but wait."

She still couldn't shake the idea that there must have been a phone call, if someone had come there for Karen to tell her so. "Professor, did you try calling her home in the States? Do you have the number of her parents in her records?"

He frowned. "No, I hadn't thought of that. We could find out that way if there really was an emergency back there, couldn't we?"

She saw Martin shake his head. "She told me her mother was dead and she and her father didn't get along well. I'd be surprised if she put his phone number in her records."

Windle said, "Well, I'll have to check and see. I'm still convinced there wasn't any such call last night."

"Then who was this man she left with?" Ann asked. "You didn't see him, Martin?"

"No, I didn't. If only I had gotten up at the beginning to take a look at him—but it never occurred to me to be suspicious then. After Karen went out I did get up and look down the hall as she left, but he had already gone downstairs."

"What did he sound like?"

"Just an ordinary man's voice speaking English. He didn't have much of an accent, as far as I could tell. I can't even be sure about the age, except that I'm sure he was older than I am. Maybe about the professor's age." Except that it couldn't have been the professor, of course, or both Martin and Karen would have recognized his voice.

Something occurred to her. "How could this man have known where Karen's room was? He did seem to know where to find her?"

Martin nodded emphatically. "Yes, I have wondered about that too. It may have been that he just knocked on the first door he came to. Hers was the first one from the stairs."

"You never lock the outer door downstairs?" she asked him.

He threw up his arms. "But why? Who would expect robbers up here?" He was evidently more upset than she had realized.

Windle said, "It does seem odd that the man went directly to Karen's room. Unless it was someone who knew her well, and that hardly seems likely. She very seldom went into the village."

"It may be," Martin said thoughtfully, "that there is another way to explain it. You see, all our rooms face on the courtyard." He pointed up at the second floor of the wing adjoining the tower. "Possibly if the man had been down at the *Gasthaus* last night—there was a big crowd there —he would have known that Karen hadn't been out with us. So maybe he walked up just after we left and saw our bedroom lights as he was coming up the road, and just came up and went to the only room that was dark when he first saw them." But he shrugged. "That doesn't explain it, though. No, he couldn't have seen those lights from very far down the road, and I had been in bed for a long time before I heard the knocking on her door. It would have to mean that he was"—his eyes widened a little—"was waiting here somewhere in the courtyard after we got back, waiting for us to turn out our lights and go to sleep. No, that doesn't make sense, either."

Now Eddie and George began chattering to one another in Spanish, and they moved away together toward the tower, still talking. "I don't think they really understand any of this," Martin said, watching them go.

"Neither do I," Windle said unhappily. "If there was no phone call, and I can't understand how there could have been, then she must have been—kidnaped." He seemed to have to push the last word out of his throat.

She left Professor Windle talking with Martin, who was going over the same story again. They hardly noticed her leaving. What could have happened to Karen? She supposedly went to make a phone call in Tellin, and now she was gone.

Instead of going back to the main building Ann went down the arcade of the *Unterschloss* and through the heavy wooden front door.

Karen's room was narrow and dark, with twin beds along one wall and a single window. An old-fashioned dresser and wardrobe on the other side. Like a monk's cell, Judy had written in that letter of hers. On the dresser were a brush, a jar of skin cream, a bottle of Chanel perfume. (*Not so liberated after all*, Ann thought wryly.) There was no note. So she either didn't have time to write one before she left or felt no need to.

Taking a deep breath, Ann made herself open the drawers and look inside. There were the usual clothes—blouses, jeans, underthings. One pair of panty hose. Two bathing suits, one over the back of a chair. Beside the dresser was a suitcase. Ann hefted it. It felt empty. And in the wardrobe, two dresses and a trench-type raincoat. On the floor of the wardrobe was a brown leather purse.

I've gone this far, I might as well stoop to that too. She picked up the purse and went through it. There was a passport issued in the name of Karen M. Phillips with the usual unflattering photograph. A batch of traveler's checks, $550 worth. A return ticket to Miami via Paris and New York. A wallet containing a few Austrian schillings, an ID from the University of Florida with a color photo that did Karen more justice. A Florida driver's license and a snapshot of a dark-haired young man in a T-shirt and jeans. He looked as though he might have been one of her Cuban friends, maybe the fellow she had been living with the winter before. There were also a German-English phrasebook and a pair of sunglasses. She ran her hand along the bottom of the inner lining but found nothing else, not even a gum wrapper.

There was the sound of tires on the gravel in the courtyard. Karen coming back? Ann could imagine her getting out of the van with a crooked smile and explaining nonchalantly that she had decided to drive down to Zirl to place the return call home and got tied up, had trouble reaching the number. . . . She went to the window. Leaning to one side, she made out the top of the black Mercedes. Alex got out, looking very square-jawed and solemn, and walked toward the two men who were still standing outside the tower. He didn't look up.

It made sense that he would come here. News travels fast in any small village and the professor's urgent questions earlier that morning must have aroused a lot of talk. He would be asking Windle now if there was anything he could do, and he would be trying to find out exactly what had happened.

The easy, self-assured way Alex had crossed the yard reminded her pointedly of the first glimpse she had had of him at the roadside café on the way to Tellin. Her first suspicions of him then had been entirely wiped away. She wondered now how she could ever have believed that he might have meant her any harm. During the bad times with her former husband, Ann had sometimes doubted that she could ever allow herself to trust any other man again. But she did trust Alex, completely.

She watched him talking to Windle and Martin, saw him nod occasionally in agreement as he listened carefully to first one and then the other.

She put the purse back into the wardrobe where she had found it and shut the door. One more quick look around. She peeked under the beds.

Nothing but a pair of sandals. She opened the suitcase and confirmed that it was, in fact, empty. She went out and down the stairs.

THIRTEEN

Professor Windle and Martin were just walking inside the tower but Alex wasn't with them now. His car was still in the courtyard, however.

He was waiting for her at the foot of the stairs in the great hall. He seemed relieved to see her. "I was just about to send someone up to find you," he said.

"I was in Karen's room." Feeling a need to explain, she added, "I hoped she had left a note—or some indication where she had gone. But there wasn't anything."

"Is there someplace we can talk? Somewhere a little more intimate than this?" He raised his hand ironically at the huge room. Behind the open door on the near wall they could hear someone rattling pans in the kitchen.

They decided to go outside and sit in his car where they could be sure no one would overhear. He opened the door for her and then got in on the other side.

"They told you about Karen," she said.

"Yes, I talked to Windle earlier when he was looking for her in the village."

"She was supposed to have gotten that phone call."

"Windle was right about that. There wasn't any phone call. I called Innsbruck just before I came up here. The airports have been checked. She hasn't taken a plane out, that much is certain. And the roads out of Tellin are being watched, besides. There has been no sign of the van or of Karen. They're certain she is still in this area somewhere."

"The police told Windle they couldn't investigate for twenty-four hours."

He pushed in the car lighter and took out a cigarette. "They had to say that because they didn't want him to know it wasn't a routine case."

"You really shouldn't smoke," she said gently.

"I know." He lit it anyway.

"Do you have any idea what has happened to her?"

"It's puzzling. From what they told me this morning she had become one of the prime suspects. Until this."

She wheeled around. "Was she? Why?"

"She had a few blank spots in her background. They know that she did go to a university in Florida. That checks out. And she spent a couple of years at a well-known girls' prep school in the East. But before that her background is a kind of tantalizing blur. And then there's the fact that when she applied for the job with Omega she lied about taking scuba diving training at the university she went to. There's nothing in her records there about that."

Something stirred in Ann's memory. "I think I can explain that. She told me a boy friend of hers in Florida taught her how to dive. I think she said she went to high school with him, or something. Maybe she lied on her application because she didn't want Omega to know she hadn't had any formal training. She might have been afraid they wouldn't take her without it."

He considered it. "That could be."

Then it suddenly occurred to her. "How did the police know what she had put down on her application?"

"I assume that after they got that anonymous tip they put a mail cover on everything going to and from Professor Windle."

Ann winced at that. She didn't like it, even though it *was* something that might help find her sister's murderer.

"I know," Alex said softly. "The things people do in the name of a good cause."

She looked out the windshield at the deserted courtyard. "I suppose I shouldn't be shocked. I just finished going through Karen's belongings. I actually searched her room. If anyone had ever told me that I—" She imagined a written form with the question: "Would you enter a strange girl's room and go through her purse?" The imaginary pencil took the multiple choice at the top and made a firm check mark beside it: *Never.* But she had done it.

"You had good reason." He sighed. "I was about to go there myself, as a matter of fact. What did you find?"

"Not much of anything. Her clothes and purse were still there. It looked as though she wasn't planning to leave. She didn't even take her wallet or her traveler's checks. Or her return ticket to the States. There wasn't anything out of the ordinary except that—" She paused.

"Except what?"

"Well, you may think it's silly, but she had the neatest purse I ever saw." She saw him smile. "I mean, women usually accumulate a lot of odds and ends in the bottom of purses. You know, hairpins and pennies and little scraps of paper. Hers was as clean as—well, it was as if someone had dumped everything out and then put just the essentials back in it."

"Interesting. Did you find anything in the wastebasket?"

She thought back. "There wasn't one in the room. Maybe she had just cleaned her purse out recently, though. It probably doesn't mean anything."

"It might, though. Was the purse new or old?"

"Old, I think. Why?"

"If she were coming here with a false identity, she might have wanted to make sure nothing in her belongings could give her away. A theater ticket stub, for instance. She might have emptied out her purse then. Or you could look at it another way. Someone else may have emptied it out, looking for something." He said, "Was there anything else?"

"Just an empty suitcase. Clothes. Alex, do you think she'll come back?"

"I don't know. The police are looking for her right now. Max is out driving the back roads too, looking for the van. It may turn up soon."

She had been watching his face. "But you don't expect that she'll be found then too, do you?"

"I'm only guessing, but somehow I don't. One thing that bothers me is that there's no way to be positive that Martin is telling the truth about what happened last night."

That hadn't occurred to her, but now that she thought of it . . . "There's only his word, isn't there? No one else heard that man. So it might not have been that way at all. But, Alex, he did sound very convincing when he told me about it just now. I think he was really upset, don't you? Even though he didn't seem, before this, to like Karen very much."

"Yes, he did seem upset about it. There's no good reason, actually, to doubt what he said."

"Alex, there's something else I don't understand. That man knew where her room was."

"I know. Martin was just telling me his theories about that."

He ground out the cigarette in the ashtray. "Assuming that Martin was telling the truth, the question is who the man was. Why would he want to kidnap Karen? According to Martin's story she didn't know him. Or pretended not to know him. That had occurred to me too, but—"

"Pretended? Then you think she might have *arranged* for him to come here for her? But why?" Then it came to her. "Alex, you don't think she and that man were working together and planned to go get the gold last night? And that's why they took the van?"

He shook his head. "At first I believed that was a possibility. But why would they call attention to themselves this way? It wouldn't make sense to arrange a fake kidnaping. Her disappearance brings the police into it, and that's exactly what they wouldn't want if they were trying to get out of the country unnoticed. No, I can't believe they hatched this up together."

"I see what you mean. If she wanted to leave here, it would have been much better for her to find some excuse."

"Any little disagreement with Windle would have been enough to allow her to leave without arousing suspicion. Or a fake illness. She could have simply left a note and taken off. Anything would have been smarter than this disappearing act, because Windle immediately called the police."

There seemed to be only one conclusion. "Then she was really kidnaped?" She found it almost as difficult to say as Windle had.

"Yes."

"So it really is a question of who that man was. He deliberately lured her away from here. . . . Do you suppose that *he* might be working for Krueger and wanted Karen out of the way for some reason?"

"That seems to be the only possibility left. But then there's the same problem. Why would anyone working for Krueger pull a kidnaping? He couldn't want the police all over the area looking for them both. Not if he wants to get away with that gold."

"Alex, there's something else. I almost forgot. Karen told me yesterday that she thinks Eddie is from Argentina." She repeated the story about the language mix-up that Karen had told her.

"Eddie. That would be Eduardo Alvarez."

"Yes."

"I don't know. As I remember, Kurt told me a while back that they had found a brother of the Alvarez boy in Santiago who said he was in fact here in Tellin."

"So he must be who he says he is."

"Apparently."

"Unless his 'brother' is a phony too. I don't suppose that's very likely, is it?"

"No."

She was disappointed that her information hadn't been of some use to him. "So Karen must have been mistaken about that. Or lying. Or maybe Martin was lying to *her*."

He was thinking about something else. "There could have been a reason for Karen telling you that story," he said quietly.

"Which is?"

"If she's involved with Krueger, she might have wanted to see how you would react when she mentioned Argentina."

Oh God, she thought. That hadn't entered her mind.

He saw how she looked. "It was just a thought. And probably wrong."

She understood. "If she were the one, there'd be no reason for Krueger to have someone kidnap *her.*"

He frowned and flexed his hand against the steering wheel. "No. Somehow it just doesn't add up. Any of it."

"Maybe she will turn up yet." He didn't look hopeful at all. "Alex, do you suppose Karen is dead?"

"I don't know."

She studied his face. "What are you thinking? Please tell me."

"I'm thinking that Krueger can't be happy about what is happening here. If he is behind Karen's disappearance, it must have been an act of desperation. Or possibly a wild diversion of some kind." He looked at her. "And I'm thinking that you shouldn't be staying here."

"You never give up, do you?"

"Neither do you, apparently."

But she knew he was right. Karen's sudden disappearance had made her realize that she was trying to deal with a situation she didn't understand at all. And she had reached the point where she felt that she had done all that she could do. She said, "All right. I'll make plans to leave later today and move down to the village." She didn't want to leave Tellin until she had some answers. She could stay at the inn, or with the Bruners.

He said that he would be busy most of the day

helping the police look for Karen, but they would have dinner together.

After Alex drove away she went inside to see the baroness, told her she would be leaving Schloss Adler, and settled her account. The baroness raised her eyebrows a little at the news, but said nothing. Nothing that is, until Ann asked her if she had been told about Karen's disappearance. Then she remarked quite acidly, "Americans are *so* unpredictable."

Good-by and good luck, Ann thought as she left the room.

She went out the front door and crossed the courtyard. There was one more thing she wanted to do before she left. She went into the tower looking for Martin. She wanted to get to the bottom of that story about Eddie. If Martin were to say that he had never told Karen he thought Eddie might be from Argentina, then she could be certain that one or the other was telling a lie. On the other hand, if he confirmed what Karen had said, then she would know that Karen *hadn't* made it up as a sort of trap, in order to see her reaction. She wanted very much to know the answer to that.

The tower workroom was empty. She didn't see anyone on the stairs. She went back outside and walked to the edge of the courtyard where the stone steps led down to the beach. The dinghy was on the water, halfway out to the boat. She could see Windle in the bow, and Martin with his back to her, operating the electric motor.

They must have had some work to do out on the boat. There was no telling when they would be back. She decided to go to her room and wait.

As she was crossing the foyer she ran into Eddie and George, who were evidently on their way from the dining room. She remembered that none of them had had breakfast that morning.

They were both wearing nylon shirts, a kind she thought had gone out of style many years before. It was difficult to keep from thinking of them as Tweedledee and Tweedledum again. She decided to try something. She caught up with them as they were going out and tried to engage them in conversation.

"George, Eduardo. How are you?"

They looked at her, half grinning, and then at each other. "*¿Cómo?*" It was Eddie who had spoken up.

"That's right. *¿Cómo está?*" How had that phrase come to her mind? she wondered. Something from an old movie, probably.

"*Bien. Muy bien,*" they said together.

"You don't speak a little English, do you? Either of you?"

Puzzled looks. "*¿Inglés?*" George said finally. He shook his head dolefully. "*No hablamos inglés. Lo siento.*"

Now what? "Karen," she said. "*¿No está—?*" She couldn't think of how to say "She isn't back?" She motioned toward the courtyard. "No?"

They glanced at one another again, then back

at her. Looking serious now. "No," Eddie said. "Karen *no está aquí.*"

She focused her attention on Eddie. He looked very friendly and eager to please, despite the language problem. Why not go ahead and ask him? If she could only think of a way to do it. The shortest route is the most direct. . . . "Eduardo. From Argentina, no?"

He looked blank. "Argentina? No, no. Chile." He seemed amused by her little mistake. A very good act, if it wasn't the truth.

Just to even things up, she turned to George, who had suddenly turned sullen again. "And George? Panama, no?"

"*Sí.*"

She wished she could have at least asked him about that day Judy was killed. He had been in the water, the first one to reach the chamber. But of course it was impossible. She let them go.

She went up to her room. She made sure her passport was still in the pocket of the suitcase. She would be needing that soon. She couldn't stand to wait there for Martin to return from the boat. She decided to drive into the village to see if there had been any news about Karen. She would come back later to talk to Martin and pack her bags. She went downstairs and out into the courtyard again.

Just before she got into her car she noticed Eddie and George standing outside the tower talking. Eddie saw her and waved good-by. She started the motor.

As soon as she reached the first switchback on the road down she knew she was in trouble.

FOURTEEN

With the car in second gear, she put her foot on the brake and felt the sickening sensation that she had heard about but had never understood. The pedal sank without resistance straight into the floorboard. There was nothing there. The brakes had ceased to exist. She pumped it again. Nothing. And again. Nothing. She barely made the turn. On one side was a solid wall of rock, on the other a precipitous drop of several hundred feet. Despite her taking her foot off the accelerator the car gathered momentum as it moved downward toward the valley.

The next sharp curve was coming up fast. An elemental part of her brain took over, directing her movements without her being able to think about it. She threw in the clutch and tried to get the engine into first gear. There was a gnashing of metal as she yanked the shift repeatedly, trying to force it into position. The engine was turning over too fast. It refused to give.

She could see the curve coming closer, and she was headed straight for the edge. She jerked the wheel sharply to the right.

It was remarkable how vivid everything was.

The numbers on the speedometer seemed to call themselves out loud, they were so clear to her. Her terrified mind fantasized newspaper headlines: A Second Daughter of Hamilton Family Killed in Austria. Except that it wouldn't be a headline, even in the hometown paper. On the obituary page. *Why didn't I listen?*

A spume of gravel flew from under the front wheel and sprayed out over the edge. She felt an impact as the fender scraped against the jutting stone wall. The car shuddered away with an explosive sound and the rear wheels skidded violently for the brink. She wrestled it back and somehow made the curve.

But it wasn't over. She remembered finally to turn off the ignition, but it seemed to have little effect as the car hurtled headlong down the slope. The village bridge came indistinctly into her line of vision, another car crossing it, moving in her direction. Coming head on. She made a last frantic attempt to pump the brakes, then swerved wildly off the road. She was bounced around violently as the car went across the hayfield. There was only one way to avoid plunging into the stream of water under the bridge and she barely had time to do it. She turned the wheel hard in the direction of the hayracks.

She neither heard the sound of the crash nor felt the impact. The engine sputtered and died, leaving an anticlimactic silence. When she opened her eyes a few moments later she saw hay packed tight against the windshield and sticking through the open window on the passenger side. She put her head down against the steering wheel.

Before long she heard the sound of someone running toward her. Then the hay began to be pulled aside. Light came in. There were two young boys in lederhosen at her window. And then she saw Alex running up behind them.

He opened the door and touched her shoulder as though he were afraid she might break. "Ann, are you all right?"

Her forehead was throbbing. "For someone," she began, then paused in astonishment that her voice was shaking like an old woman's, "for someone who just drove down a mountain with no brakes, I'm fine."

"Sit still for a minute. Make sure you're okay." He kept his hand on her shoulder. The young boys had crawled up the haystack and were looking at the front of the car. The hood had popped open in the impact and was stuffed with hay. One boy said something to the other, and Alex laughed in relief and told her, "They never saw a car that ate hay before."

He sat her down in a rocking chair in his living room, got a small fire started in the old-fashioned stove, and when he saw that she was still trembling, brought her a rough wool blanket to spread over her lap. Without a word he went into the little kitchen. Before very long he came out with two steaming mugs of cocoa.

"Go ahead and say it," she told him, as he handed her the cup. *I told you so.*

"No, I'm not going to." After she took a sip he said, "I was going to make coffee but then I decided you shouldn't have a stimulant right now. I was just on my way up into the mountains when I saw you careening down the castle road. I thought at first you were trying to see if a Volkswagen could fly."

She managed a pale smile. "For a while there, I wished that car did have wings."

He slouched down in a chair opposite her, holding his cup without drinking. "How do you feel?"

She rubbed her forehead. "There's a bump coming. I must have hit my head on the steering wheel. But I'm okay. Just give me a few minutes to get my nerves back together."

"There's no doctor in the village. Would you let me drive you down to Zirl?"

"No, don't be silly. I'm perfectly all right. Not a scratch, as you can see." Actually she did feel much better now. But it was beginning to sink in how close a call it had been. "It was like trying to drive down a crooked tightrope. With no net. Something was wrong with the brakes, Alex. They just weren't there. I was supposed to be another accidental death."

He didn't say anything.

"Who could have done it? I don't know why anyone would think I knew enough to be a threat to them. Maybe they didn't expect me to die. Maybe it was meant to be a warning to stay away." She realized she was talking too fast but she couldn't help it.

His voice was very quiet and solemn. "I shouldn't have let you go up there."

"You couldn't have stopped me. Short of tying me up."

"Then I should have tied you up."

He watched her carefully while she took a long time finishing the cocoa. She had calmed down considerably by that time. He stood up and held out his hands to her and brought her to her feet. He felt the bump. "That hurt?"

"No."

He looked at her face. "Come on, I want you to go upstairs and lie down. If you're sure you're all right, I want to go take a look at your car."

After he left her she slipped off her shoes and leaned back on the bed, pulling the blanket over her. She felt a little drowsy because of the cocoa, but she knew she wouldn't be able to go to sleep. She heard him go out and close the front door, and after a while she sat up again. She glanced around her.

She liked his room. It was like him. Interesting, comfortable, neat. . . . Hanging on the wall was an old stringed instrument like a dulcimer and one of those Austrian carved masks, the kind that was used in festivals. A print of Albrecht Dürer's "Lion." An overstuffed chair with lamp and table for reading. Lots of books on the table and piled on the floor by the chair. Across from the bed was a record player and a stack of albums. She could see a Mozart horn concerto and a Grieg. And a

Miles Davis album, which surprised her. She hadn't expected he might like jazz. There was a lot about him she didn't know.

Feeling restless, not wanting to dwell on what had just happened to her, she got up and put on the Mozart, turned it down. She went over to the chair and picked up a book. *The Rise and Fall of the Third Reich.* His homework, she thought. She read some of the other titles. There were more books about the Nazi era, and a wide selection on other subjects as well. Books on biology, essays on the Renaissance, a John Updike novel—that was another surprise—*The Brothers Karamazov*, a book on the European economy. She read one of the biology titles: *Die Hochgebergseen der Alpen.*

She had always liked the way the German language made compound words out of everything, jamming adjectives and nouns together like boxcars in a train. But the sentence structure still threw her sometimes, the way the subject of the sentence was often left out until the very end. She remembered how Mark Twain or someone had said, "Once the German language gets hold of a cat, it's good-by cat." She opened the book and tried to read about the Highmountainlakes of the Alps, but the text was too technical for her.

She scanned some of the other titles. She picked up a book called *Hitler's Men* and sat down with it, turned casually to the index. There was only one listing under Krueger, Heinrich: a footnote that said he had disappeared from his command post in the Tirol in the spring of 1945. He was a minor figure, obviously. But still alive, apparently, like a peculiarly resistant malignancy.

When Alex came back he told her, "The car isn't as badly damaged as I thought. A smashed fender and a tire blown out. The first thing caused the second, evidently."

"That must have been the loud bang I heard," she said. "I think that was what almost sent me over the edge."

"It may have saved your life once you got the car under control again. That one flat tire probably slowed you down a bit." He pulled up a footstool and sat down in front of her. "How are you feeling?"

"I'm fine now. Did you find out what happened to my brakes?"

He took a deep breath. "It's not good news, I'm afraid. Someone yanked the brake line loose. I had Erik Steub, the man who worked on your car before, come out and look at it. He took your car over to his place to make the repairs."

"Someone at the castle tried to kill me."

"Yes, it must have been, Ann."

"Is there any way of telling when it was done?"

"It could have been any time since you last used the car. The brake fluid would have drained out immediately, and after that, no brakes."

There wasn't much to go on. She hadn't used the car since she arrived at the castle. "It must have been done at night, don't you think?" Anyone living there would have had the opportunity.

"Probably."

"Last night, maybe?" She remembered what had happened the night before. "The same man who came to get Karen might have done it?"

"I don't know, Ann. I wish I did."

"This person, whoever it is—Alex, it looks as though he will kill anybody who even threatens to get in the way." She said it quite calmly. "I don't even know anything that could be a threat to him, except that Judy was murdered."

"Someone panicked, maybe, thinking you knew more. Or possibly we have a real psychotic on our hands. It could be that Krueger didn't choose too wisely when he picked someone to send here."

She was still holding the book about Hitler's men in her lap. She glanced down at it. "Were they all like that, psychopaths of some kind? The Nazis, I mean."

"No. The West has made a mistake to dismiss them that way."

"Why a mistake?"

"Because they weren't. Not all of them. It would be nice if it were that easy. Hitler didn't come to power, after all, by promising war and genocide. He promised peace and bread and justice, all those nice abstractions ordinary politicians like to use."

She tried to imagine Hitler giving a conventional political speech, and couldn't, although she remembered hearing that he was a charismatic political speaker. "But most politicians don't—aren't like Hitler. So how could you be sure?"

"Whether he'll turn out to be another Hitler?

I'm not certain. I think it has to do with ends and means." He got up abruptly and changed the subject. "You aren't going back to Schloss Adler." It wasn't a question or a command, just a flat statement of fact.

"No." Her voice was very still. It was all beginning to catch up to her. "You know, when I was driving down that hill I wished I had never come here."

He leaned down to kiss her lightly on the mouth. "In spite of everything, I'm glad you did."

Later, after he had brought her up some hot soup, which she ate, and after she kept insisting she was perfectly all right, he said he would take her to Frau Bruner's to stay while he completed what he was starting to do earlier, which was to go help Max look for Karen and the missing van. They were about to walk out the front door when he turned her toward him gently and put his arms around her. He kissed her for a long time. And she kissed him, letting her arms go around his neck as though they had always belonged there.

Finally she pulled away and smiled at him. His face, which had at first seemed like a travel folder Austrian's to her, had revealed its uniqueness hour by hour. The strong bones of the cheeks and jaws, the tiny scar that made a brief irregularity in the left eyebrow, the way his eyes changed from light to darker blue against different backgrounds. With the dark shirt he had on now, his eyes were almost gray as they looked back at her solemnly.

I love you. She was astonished at how readily the words had formed in her mind. "One of these days," he said, "I want to have a serious talk with you."

She thought it was a funny thing for him to say. "That's all we have been having. Serious talks."

"This is something else."

She raised her hand to touch the tiny scar on his brow. "Where did that come from?"

"I fell mountain climbing when I was a kid."

She looked at it again. "I thought maybe your violin teacher hit you with the metronome."

He laughed. "No, he probably had a nervous breakdown instead, poor man."

He kissed her again and opened the door.

FIFTEEN

They could hear a radio playing inside before Frau Bruner answered the door. Her startled look, when she appeared, went from him to Ann and back to Alex as though she had been expecting anyone else but them. "Alex?"

"I've brought Ann back to stay with you, Elsa. Is that all right?"

Her wide eyes shifted back to Ann. "Of course. Yes." But she didn't, Ann thought, sound too sure.

"Do you know where Max is?"

"He went into the mountains toward the Bender gorge. He said he would be late getting back."

Alex nodded to indicate he understood where that was. "I'm going up there myself in a few minutes. I have to go back to the inn first to make a phone call. I want you to take good care of Ann while I'm gone, all right? She was in a car wreck earlier today."

Elsa's face softened. "Oh, I am sorry. Of course, Alex." She opened the door wider for Ann to come in.

Alex looked at her. "I'll be back as soon as I can." And then he left.

Frau Bruner closed the door behind her and turned around, one rough hand clenching the other. "I don't think—" she began.

She obviously hadn't been prepared to have a guest that day. "If it's going to be inconvenient for you, Mrs. Bruner?"

"No, no. You are Alex's friend. We have room."

"If you're busy, I could go upstairs now and let you get on with whatever you were doing. I don't want to get in your way."

"No, no," the woman said, too quickly. Something was wrong. "Let us go into the kitchen for a minute."

Somewhere upstairs the radio stopped playing.

When they got to the kitchen Frau Bruner said, "I won't be able to let you use the same room you had before. My daughter is here now. Ordinarily she stays at the *alm* house in summer, but when Max saw her the other day when he was up there, he said she didn't look well to him, and so he went back yesterday and brought her home. But it's all right, we have another room."

It was obvious from the worried expression on her face, however, that it wasn't all right, and that she had agreed to let Ann stay only because she didn't want to refuse her friend Alex. Freida was here, and Freida didn't like to be around strangers. Ann was about to protest when Frau Bruner said, "There is another bed downstairs. Come and I will show you."

They passed through the hall, Frau Bruner in front, and a movement at the top of the stairs made Ann glance up. A door had opened and she could see Freida standing there peering out uncertainly, her face in shadow. At the *alm* her impression of Freida had been that she was much older than her age. Now she seemed too young. Her heavy face had the softness of a girl who hadn't grown up. When Ann glanced up, Freida drew back quickly and shut the door. The meaning was unmistakable. As long as Ann was there, Freida would stay in her room.

She caught up with Frau Bruner. "This isn't necessary, really. I don't want to inconvenience you. There's an inn in the village. I'll go there."

The older woman protested, but her face was filled with obvious relief. "If you're sure. You're sure that would be all right?" Ann told her that it would be and thanked her for the offer to stay, just the same.

She decided to walk back into the village to try to find Alex or at least leave a note at his cottage before going to the inn. It was only a short walk into the center of the village.

As soon as she started out down the road, however, she began feeling dizzy.

It was because of the accident, she realized. *The attempted murder, you mean,* a sardonic part of her brain reminded her. She stood still for a long moment staring at the gravel at her feet. When the dizziness went away, she began to walk down the road again, more slowly this time. Luckily she didn't have anything to carry but her shoulder bag. She was thankful she hadn't brought a suitcase along.

She passed the white house with the big barn Alex had mentioned and saw her car. The barn door faced her, and the Volkswagen was inside driven partway up a homemade ramp. She decided to cross the road and go over there, both to find out about the car and to give herself a break in the walk.

The small garage had been squeezed into one end of the barn so as to use every available inch of space. A brown horse eyed her dolefully over a wooden partition behind the car. The smell of oil and gasoline mixed strangely with the aroma of fresh hay. She went far enough inside to see the damaged part of the auto. The right fender and door were badly dented. The exploded tire had been taken off the front wheel and was lying on the dirt floor.

She went to the house and knocked on the back door. In a few moments a wiry middle-aged man appeared, and when she told him who she was he came outside at once. She thought that the way he looked at her was odd. His eyes searched her face as though he were trying to find the answer to some puzzle there.

They went over to the car and he pointed out the damage. "The brakes I have fixed," he said carefully, "but the left front wheel is badly out of line. I can repair that tomorrow, if you like. I have a spare tire that will do. The broken headlight, I don't have the right parts for that. I could send off, but how long it will take I don't know."

"No, that's all right. I could have that taken care of later. I never like to drive at night, anyway, and I'm not sure how much longer I'll be staying. Would it be all right if I just left the car here for a while?"

"Of course, fräulein." He was still looking at her in a peculiar way.

"I can pay you now."

He shook his head. "But there is no need. Alex Schuler has already taken care of it."

"Then I'll have to pay him. How much was it, please?" He named what seemed a fair price. She was sure the auto rental's insurance company would reimburse her eventually.

"Alex told me," she said, "that someone had tampered with my brakes."

"That is definitely my opinion, yes."

Without real conviction she added, "There's no

chance it could have been an accident, I suppose?"

He looked amazed, almost indignant, and she thought: *Why is he acting so strangely*? "After what happened before? I should think that it's not likely. Two similar incidents in less than a week?"

She stared back at him. "I don't understand what you mean."

He looked even more surprised at that. "I'll show you." She followed him to the other side of the car, where he took down a long band of black rubber dangling from a nail on the wall. It was her broken fan belt, he said. He brought it outside into the light so she could see it clearly. He showed her how the end was frayed about a quarter of the way through. The rest of the edge was flat and smooth.

"Someone cut this partway so that it would fail in a short time, after a few miles. A rather harmless prank, I suppose, if you stopped before the motor was damaged, but the brakes—that is a serious matter."

The scene flashed into her mind like a bulletin: Alex standing up behind her car and patting the pocket of his jacket as though he had dropped something, then walking past her toward the roadside café.

The man's eyes moved from the damaged belt in his hand to her stunned face. Now he had found a different mystery to puzzle over. "Do you mean to say," he said in amazement, "that

Alex Schuler did not tell you I had shown him this belt?"

When he opened his door and smiled at her in surprise she brushed past his shoulder, then wheeled around to face him. He had on the same windbreaker jacket he had worn that day; he was just getting ready to go out.

"Ann? What's wrong?" He looked genuinely concerned. "I thought you—"

She tried her best to keep her voice under control. "Why didn't you tell me you had cut the fan belt on my car?"

"I was planning to." Just like that.

So he wasn't even embarrassed. That made her more furious still. "Were you really?" she said acidly.

"Ann, you are right to be angry." He knew better than to try to move closer to her or to reach out to touch her. "Will you at least let me explain?"

She gave him a look that she expected him to flinch from, but he continued to meet her eyes directly and calmly. "It might be interesting to hear you try. But I only want to know one thing. You knew ahead of time that I was coming to Tellin, didn't you? You must have known."

"Yes."

"How could you have possibly known that?"

"It all happened very quickly. On the day you left New York the policeman you had called earlier for an explanation about Judy tried to call you back. He wanted to talk to you again for some reason, I don't know what. Your roommate answered the phone. She told him where you had gone, and she must have said something to indicate that you were coming here to do some investigating of your own. The policeman became alarmed because he didn't know what you might do. You see, after your sister's death Kurt Reinhardt had gone to that policeman and told him he wanted to know everything he uncovered concerning Judy's death, and he told him why. He told him about Krueger. The authorities in Vienna had wanted to keep the local police out of it, but at that point, with them going to Schloss Adler to investigate what everyone believed to be a diving accident, Kurt felt he had no choice but to tell them. And so when the police officer learned you were on your way here, he called Kurt to let him know that something unexpected had developed."

"And then Kurt called you, I suppose."

"Yes. He made it sound extremely urgent. If you were distraught or overly emotional—he had no way of knowing you as you are, Ann. He was afraid to let me tell you the situation before I knew you could be relied on, and so was I."

"I see."

"And I didn't think I could let you walk into a situation like that *without* knowing—"

"So you had to come up with a way of finding out what I planned to do and how much you should tell me very quickly, before I reached the castle." Her voice was brittle, like ice. She was still trying to stare him down, but he wouldn't look away.

"Look, don't misunderstand what I am saying to you. I am not trying to justify what I did. I don't believe I can. I regretted it the minute I did it. Especially after I had met you."

"Of course. I understand completely. And so you went out and . . . and what? Watched for my car?"

He nodded grimly. "Yes, they found out the flight you arrived on and the make and license number of the auto you had rented. I was in Vienna that weekend and Kurt called me at my apartment and explained the situation. I drove back and waited for you near Innsbruck and followed you until you stopped at the café."

"And cut the fan belt so that you could come along later and—and *rescue* me and drive me in to the village. Sir Galahad rescuing the damsel in distress. Trying to pump me for information along the way."

"Go ahead, say it all. It's not worse than what I've said to myself."

She was clutching the strap of her shoulder bag so hard that her nails were digging into her hand. "Didn't it even occur to you to do the simple thing—just introduce yourself to me at the coffee house?"

"And tell you the whole story about Krueger and the rest? And when I told you I knew who you were and why you had come to Austria, that I had followed you—"

It would have frightened her, all right. And she couldn't be certain now how she might have reacted. Certainly he had no way of knowing what she might have done. But her anger didn't diminish. "Very well, I'll give you that much. You were under pressure. You thought you had to act quickly. And you didn't know what I might do. That's not what makes me angry. Afterwards, after you decided you could risk telling me what was going on, why couldn't you have told me the rest then? I thought I could trust you, Alex. I thought we were friends." She realized she sounded bitter.

"Will you try to believe me when I tell you the reason? After I became convinced that I couldn't persuade you not to go to Schloss Adler, I wanted you to be able to trust me enough to rely on me if you needed help. And I knew you wouldn't do that if I told you the truth then." Probably that was true too, she realized.

"I have wanted to tell you before this," he said earnestly. "I was going to tell you that day at the *alm.* I wanted to apologize. But then we saw Freida."

She remembered how he had seemed troubled that day about the methods the police had been using, how he said he had been pressured into doing things he didn't like. That was just before Freida appeared, and she had felt then that he was about to say something else about it. And yet this somehow didn't help matters. For some reason it carried no weight. She felt he had betrayed her.

She said angrily, "It was very convenient for you that Freida did happen to come out of the house, wasn't it? How do I know you ever intended to tell me at all?"

He sighed heavily. "You learned about it from Erik Steub, I imagine. It was inevitable that you would talk to him sooner or later, when you went to pick up your car. Do you think that if I wanted to keep this from you I wouldn't have warned him not to say anything to you?"

Part of her mind told her that what he said made sense, that if he had planned to deceive her indefinitely he would have tried to cover his tracks. A painful thought came to her: That's what Don would have done. Don had always carefully arranged for one of his friends to cover for him when he was seeing someone else, yet it seemed she always found out anyway. And the lies only magnified the hurt.

He was eying her speculatively. "Ann, it puzzles me. I knew you would be angry, and I admit I deserve it. But you are more upset now than you were when you first walked through that door. It makes me think there's something else—"

"You lied to me, Alex," she exploded. "I hate lies."

Almost immediately he said, "Because your former husband lied to you, is that it?" His voice was gentle but she turned away abruptly, stung. Now he did come over to her and put his hands on her shoulders. "Ann, please believe me, I—Ann, look at me."

But she wouldn't. He had come too close to the

truth, opened the old wounds. Her voice fully in control now, she said, "When I first came to Tellin you tried your best to persuade me to leave. Well, I'm leaving. Just as soon as I possibly can." She twisted away from him and ran out the door, slamming it behind her.

She would go to Innsbruck, she decided, and talk to the police. She would insist that they re-open their investigation of Judy's death, and she would stay there if necessary to see that they did it. She would have them talk to Professor Win-dle, and she would tell them how she had almost been killed herself. At the moment she couldn't think further ahead than that. She simply wanted to leave Tellin and try to forget she had ever come there.

The narrow lane from Alex's cottage to the main street of the village was fairly steep. In her haste to get away before he decided to come after her she found she was hurrying too much and getting out of breath. She stopped for a moment to rest with her hand on a wooden fence before going on. There were people passing in the street up ahead. An old woman and a young boy. They looked blurred when she glanced up. *I have to take my time.*

Before she reached the inn she had decided she would have to hire another car if she could. She didn't want to wait until the other one was repaired; she would have someone come back for it later.

She located the manager of the inn, a stocky man with a cherub's face who understood little English. With difficulty she made him understand what she wanted. As well as she could make out, he said he had a neighbor who owned a car who might be willing to drive her to Innsbruck, but he wasn't sure. He asked her to wait.

As she sat on the little sofa in the lobby she realized that she hadn't brought her passport with her. It was in her suitcase at the castle. Now she would have to go back there after all.

The innkeeper's neighbor was a dour man in his late fifties who indicated he would take her to Schloss Adler and then on to Innsbruck. They agreed on a price.

His rusty old car sounded as if it needed a new muffler. Ann had doubts that it would make it up the road to the castle, let alone to Innsbruck. But they arrived at Schloss Adler without any trouble, and she asked him to wait for her while she went to get her things.

It didn't take her long to throw her few clothes into the suitcase and make sure her passport was still there. She intended to leave without saying good-by to anyone. She would contact Professor Windle later and thank him for his help. She didn't care if she never saw any of the others again.

But when she came downstairs Martin was hurrying across the great hall in her direction. He looked more than ever like a boxer in a sweatshirt and jeans. His face lit up when he saw her. He met her at the bottom of the stairs. "Hello, Ann. I saw you get out of the car a moment ago. We have been worried about you."

"Have you?" She hoped her voice didn't sound too hollow, but somehow she doubted his sincerity. She recalled that she had wanted to come back there to talk to him about something. Yes, about Eddie. It didn't seem important now. How could she have expected to be able to find out what was really going on, she wondered. It was useless. She only wanted to leave.

"Ann, we heard about the accident. I went down to the village to see about you, but I couldn't find you anywhere. They said you had gone away with a friend." He looked a little hurt. "Are you all right?"

"Yes."

"Then you weren't hurt?"

"No." As a matter of fact her head was pounding. And the dizziness was coming back again. She put the suitcase down.

He looked at it. "You aren't leaving?"

"Yes, I have to go, Martin. The man in the car is waiting for me." But then she added, "I just want to sit down over there for a minute. I don't feel—" The chair three feet away suddenly looked as though it had passed behind a gauze curtain.

"Ann, *qué pasa*? What's wrong?" His voice came from the far end of a tunnel.

Martin caught her as she fell.

SIXTEEN

Patterns of light on the wall.

Meaningless blotches of light. Elongated rectangles, sunlight through a window. She turned her head. The ceiling was familiar. It was daylight. She knew she must have slept for a long time. Overnight? She recognized the bedspread. She was in her room, one of the castle's upstairs bedrooms. Someone had put her in her nightgown. Not *Martin*, she hoped. She tried to sit up but she still felt weak. Her head hurt.

There was a noise in the hall, and she guessed that it was that sound that had wakened her. There were footsteps coming closer. She tried to clear her mind enough to face whoever it might be.

Baroness Olga. She opened the door quietly and peeked in. "Ah, you are avake. Goot." When she came inside, she was almost tiptoeing. "You are feeling bedder now?"

"I suppose so." She tried to sit up again.

"Here, let me help you." The baroness came over and pulled the pillow up behind her, fluffing it in a motherly way. "There. I had been up several times to see how you are doing. Sleeping every time I looked."

"I'm sorry to put you to any trouble, Baroness von Toblen."

"Nonsense, my dear girl," she said cheerily. "It is understandable. We know, of course, about the unfortunate accident with the automobile you had yesterday. It must haff been a very frightening event for you. No vonder that you fainted, my dear. The fright vas too great."

So she must have slept all night. "It's not that," Ann replied weakly. "I hit my head on the steering wheel." She touched her forehead gingerly. The bump was still there, bigger than ever.

The baroness's dark eyes followed Ann's hand. "Oh, dear. Yes, there is a bruise."

"Yesterday was not a good day for me." She remembered walking out on Alex.

Baroness Olga brightened, pursing her lips in a smile as though she were about to let her in on a delightful secret. "But today you are lucky," she said. "An old friend of my husband's arrived for a visit just last night, and he is a doctor. I'm so happy now that he happened to come at this time. So you see, you need not concern yourself. I vill ask him to come up to see you. I'm sure it shall turn out to be nothing quite extraordinary." She meant, of course, "serious."

Ann thanked her and said that no, she didn't want breakfast sent up, she couldn't eat a thing. Perhaps some coffee. The baroness fussed with her pillow again and clucked sympathetically some more and then she went out.

Ann wanted nothing more than to get up and leave, but she knew that it wasn't possible yet.

Even lying propped up in bed, she felt sickening waves of dizziness. She could no longer tell herself that it was an emotional reaction to the strain of the past day. Something was very wrong. Something physical, not psychological.

Within a few minutes the baroness came back with the coffee. "The doctor is coming up very soon," she promised. "Professor Windle has been asking about you this morning. And one of the young men too. I don't remember—"

"Martin?"

"Yes. They ask me if they can come up to see you."

"No, please. I don't think I'm up to that. Tell them—tell them I'll try to stop by to see them before I leave."

The doctor came in. He was a tall, distinguished-looking man in his sixties, dignified and neat in a dark suit, with a trim mustache and goatee. Dr. Praeger, the baroness called him. He looked at Ann and said rather stiffly, "How do you do?" He was carrying a black bag, which he set on the dresser.

She wasn't particularly interested in the coffee now, she was more eager to have the doctor tell her what was wrong. The baroness made a move to leave, but the doctor motioned for her to stay. He took Ann's temperature, listened to her heart, looked into her eyes with one of those little flashlights. As he bent over her she noticed a sweet, not unpleasant scent on his breath. What an unusual mouthwash, she thought to herself.

"You fainted, they tell me?" His accent was softer than Olga von Toblen's.

"I was in an accident yesterday." She checked herself. "No, not an accident."

He drew back with a stern frown. "Which was it then?"

"It's not important, I suppose. I hit my head on the steering wheel."

"I see." He stared at the mark on her forehead. "You have been dizzy afterward?"

"Yes, and my head hurts very badly right now. At first there wasn't any pain or anything." She looked at him questioningly.

He shrugged slightly. "It happens that way, sometimes."

"Then you think it's a concussion?"

His eyes widened a bit. "Very likely. Very likely."

When he went back to his black bag, Olga, who had been hovering near the bed, leaned over and asked curiously, "I thought I heard you say there vas *not* an accident? You do remember you said that?" She seemed, Ann realized, to be worried that the blow may have affected her senses. She seemed to be speaking as much for the doctor as for herself: *See, there is something very wrong with this girl.*

Irritated, Ann said very distinctly, "It's a peculiarity of the English language, Baroness, that car wrecks are always called accidents, even when they aren't accidental. Even if someone deliberately drove a car into a tree, they'd still call it an accident."

Her eyebrows flew up. "Deliberately? You didn't—"

"Oh no, no." She felt very tense, and annoyed by this tiresome woman. "Actually, someone at Schloss Adler disconnected the brake line on my car and destroyed my brakes. Someone tried to kill me."

The baroness drew back. "Oh, dear." Her eyes darted nervously toward the doctor, who was still standing with his back to them going through his bag.

Ann already regretted speaking so openly. It would only bring on more questions.

But the doctor, when he turned around, didn't seem alarmed at all. He brought her two pills and handed her a glass of water from the bedside table. Doctors have heard practically everything at one time or another. "You need rest," he said. "Take these." He was very businesslike, she noticed. He didn't look at her directly. She took the pills from his hand. She remembered reading somewhere that the only thing doctors could do for a mild concussion was to see that the patient stayed in bed and got a lot of rest.

He reached into the right pocket of his suit coat and pulled out a little gold-colored piece of candy, unwrapped it and put it into his mouth. His faded blue eyes were staring off thoughtfully at the wall behind her bed. She heard him bite down on the candy.

So that was it, she thought. Butterscotch. Without any resistance, she swallowed the two tiny pills.

* * *

Sleep came quickly. A heavy possessive sleep that dragged her down into a soft black pit with steep sides. She didn't dream conventional dreams. She only felt the presence of that deep pit into which she had fallen. Motionless, without sound, a blackness so deep that it seemed to come from the middle of the earth. She didn't toss or turn. It was like lying in a coffin.

She slept for a long time. Eventually even the black pit disappeared. Hands touched her. Touching her forehead, her cheek. Someone. Picking up her wrist from under the cover, holding it between fingers and thumb. Searching for signs of life. How odd.

Her eyelids seemed pasted together. She forced them apart. Saw the blurry top of a gray head. The doctor looked at her and dropped her wrist.

"What?" she asked. She was surprised that she could speak so cogently. That one word was so meaningful. It expressed everything she had ever wanted to know. Probably it expressed all the meaning of the entire universe. She wanted to try to remember it when she was awake again. Surely the doctor would realize how brilliant it was. "What?" she asked again, and smiled.

He didn't answer. The faded blue eyes scanned her face, and he stepped back. She had trouble focusing her eyes on him as he moved across the room and then came back. He was holding something in his hand. His other arm reached behind her back and raised her up. He was trying to give her more medicine.

"No," she said weakly. She wanted to know the answer to her question. She wanted to come awake. There was something important she was supposed to do, she couldn't remember quite what it was. She only knew she didn't want to sleep any more.

But the arm against her back was insistent, wouldn't let her turn away. Her head hurt. Maybe he was right. When she was sleeping her head didn't hurt and she could rest. She needed rest. He was the doctor, wasn't he?

She took the pills. Swallowed them and took a sip of water from the offered glass. There had seemed to be more of them this time. Three? Maybe even four. She wished he hadn't given her so many. She ordinarily didn't like to take medication at all, not even aspirin, and she didn't like doctors who prescribed too many pills.

And there were other things that she did not like. The doctor didn't even wait to see that she went back to sleep. He didn't look at her again, as a matter of fact. He closed his bag and hurried out of the room as though he had someplace important to go. He certainly hadn't tried to use his best bedside manner. If she were awake when he came back she had a good mind to tell him that bedside manner was very important, that having confidence in one's doctor was half the battle. But she had a feeling he might not come back. Maybe it didn't matter, anyway. When she woke up she would be okay again and then she could leave.

She didn't fall asleep right away. She was fighting it, she knew. She wished she hadn't taken those pills. She hadn't even had a chance to ask him how she was progressing or how much longer she would have to stay in bed. And he hadn't asked her any questions about how she was feeling. He didn't act much like a doctor at all. Maybe he's not a doctor.

But that's silly. The baroness said he was a doctor, he must be a doctor. Why would she lie? The pills must be making her confused, she decided. Just a little more rest and she would be all right.

She stopped fighting and let the all-encompassing darkness fall over her.

SEVENTEEN

At that particular moment Alex Schuler thought Ann Cole was back in New York. An hour after she had left his cottage, time enough for her to cool down he hoped, he had gone to the inn to look for her. When he had opened his door just after she slammed out of there and saw her walking up that way, it hadn't surprised him. He knew she wouldn't want to go back to his friend's house after the scene they had had. He assumed she would stay at the inn overnight and pick up her car the next day. But the proprietor told him she had hired someone to take her to Innsbruck. So she was gone, he thought. He was angry with himself for not following her immediately. He wanted very badly to see her again. *This damn job*. Why had he ever agreed to it?

The next morning the missing van was discovered off a wooded road five miles into the mountains. He drove out to the location to look at it before they brought it in. There was no sign of Karen Phillips. The ground was hard, there were no tire tracks of another vehicle, and nothing inside the van to indicate where the girl might be. Except that there was a palm-sized patch of dried blood on the carpeting between the two bucket seats. A carload of policemen were there searching the surrounding area, turning over clumps of decaying leaves and poking through the undergrowth. He could tell from the aimless way they moved around that they didn't expect to find anything.

When he got back to Tellin Max Bruner had left a note on his front door. Max had been staying in close contact with one of the village women who did housework for the von Toblens. It was another way they had of keeping tabs on what was going on at the castle. The note told him that shortly before noon she had informed Max of an unusual break in routine at Schloss Adler. One of the guests, a Miss Cole, had fallen ill the day before and had been put to bed. Alex turned around and hurried back to his car.

In the foyer Baroness von Toblen told him that Ann Cole was indeed ill. Something to do with a head injury in the accident, the doctor had said. He asked to speak with the doctor. She told him he had had to leave unexpectedly. He had only stopped for a brief visit because he happened to be in the area on a trip. He had arrived very late the previous night. He was an old acquaintance of her husband's, though she herself had never met him before. He asked to see Ann. No, she said, the doctor insisted that Miss Cole should not be disturbed for any reason.

He went up anyway. The baroness trailed behind him up the stairs, protesting in a high, thin voice.

The shades had been drawn, the room was dark. He found her lying on her back in the bed. She looked very peaceful. Her skin was cold. She didn't respond when he touched her forehead and her cheeks. What had the doctor given her, he wanted to know? The baroness said she had no idea. He tried to find her pulse, growing more and more worried. Where was the doctor now? How could he leave her when she was so ill? The von Toblen woman was getting fidgety now, as though she were being accused of something.

His questions became more urgent. What was the doctor's name? What did he look like? The baroness could, at least, describe his physical appearance. About sixty-five, gray hair, blue eyes, tall . . .

Before she finished he had thrown back the covers and was lifting Ann out of the bed.

* * *

There was water running near her mouth and splashing on her face, the first sensation she was dimly aware of. She resisted the sound of it. Too loud. The water was cold. She wanted desperately to slip back. *Leave me alone.* She groaned and tried to sink down to the floor, but something was pinning her around her waist, her arms dangling down touching her bent knees.

As soon as she made a sound he turned her head roughly and forced open her mouth. Very soon she gagged. He had run his fingers to the back of her throat and pressed down hard. A stream of hot liquid rushed up her throat and gushed out her nose and mouth into the sink. He was continuing to hold her around the waist, keeping her head down.

Speaking German, he shouted to the baroness in the hall, "Bring some warm milk. And the strongest coffee you can make." He wiped Ann's face and hair with a damp cloth and hefted her upright, pulling one of her arms around his neck. "Walk with me. Come on now." Her knees were as useless as a marionette's.

He struggled with her out the door and down the hall, half carrying her until, groaning and protesting, she began to walk unsteadily beside him. He kept pulling her back and forth along the hall until Olga von Toblen came back with a tray. He made Ann drink the warm milk and then there was another unpleasant session at the bathroom sink. He washed her face again, still holding her.

"Oh, God," she moaned. "Alex." She blinked her eyes, and climbed painfully out of the pit.

"Yes."

She blinked, trying to make sure. She stared at him and her eyes refused to track, as though she were drunk. "Alex, I've been thinking," she said sleepily. "I am so sorry for the things I said to you."

He smiled. She was pretty sure she saw him smile. "Don't worry about that. My fault. Come on."

He made her walk some more in the long corridor outside her room and then gave her the coffee. She still wasn't able to walk steadily on her own, and at times her head dropped down and her eyes closed. He talked to her constantly, prodding her with questions that he insisted she answer. If she fell silent, he shouted at her. When she was able to tell him her home town and her mother's maiden name and to spell her own name without slurring her speech, he let her sit down and rest in a chair in the hall they had passed many dozen times. He brought her more coffee, and the baroness brought her a robe and helped her put it on.

Within an hour another doctor came, a real doctor this time. He looked at her and asked several pertinent questions. She was able to describe the pills she had taken the first time, and she assumed the others were the same. He nodded and his eyebrows shot up in recognition. A very strong sedative, he told her. She was lucky to be alive. An hour or two more and it might have been too late.

He checked her reflexes and looked into her eyes. The problem from the mild concussion, if that was what it was, had apparently subsided. He prescribed rest but no prolonged sleep for at least six hours.

She learned that when she had fainted into Martin's arms she had been taken directly upstairs and someone had been sent out to tell the man in the car that she wouldn't be using his services after all. They had brought her suitcase back upstairs from the great hall. And so after the doctor left, the baroness unpacked a sweater and slacks for her so she could get out of the rumpled, sour-smelling nightgown, and she lay back down on the bed. After she had combed her hair, she asked to see Alex again.

It was the first time she had seen his face clearly since he arrived much earlier that afternoon. He looked exhausted. He sat down beside the bed and took her hand. "How do you feel?"

"I'm tired. You must have walked me twenty miles." She smiled at him. "Thank you, Alex."

He shrugged it off as she knew he would. "I should have warned you that we Austrians consider walking a national pastime. I got carried away." And then seriously, "I thought you had left the country. If Max hadn't found out you were here and let me know . . ."

"What about that doctor who gave me the pills?"

"He's gone, don't worry."

"Of course he wasn't really a doctor, was he?"

"Are you sure you want to talk about this now?"

"I *have* to. I've been wondering about it ever since I woke up. I had never seen him before,

Alex. All I know is that he turned up here last night posing as a doctor. It was Heinrich Krueger, wasn't it?"

"Yes, I believe it was."

"How did he get here, I wonder?" She was calm enough, now that she was safe, to take an almost academic interest in the question.

"That's exactly what I would like to know." His voice had an edge to it. She could tell he was angry. "His picture was distributed to all the border checkpoints two months ago. Someone let him get by."

She thought about it. "He was wearing a goatee, I remember. Maybe that made a difference. I can see how it could happen, anyway. He really did look like a doctor, you know. He was so old, he looked as harmless as a cat." Then she said, "No one really expected him to come back, did they? I'm surprised he would risk it. Why would he do it, take a chance like that?"

"It was a damned reckless thing for him to do. He must have been getting reports on what was going on here. From his point of view, things must have been getting out of control—first your sister's death, then Karen's disappearance, then that attempt on your life. Maybe there was some problem in finding the gold, or arranging to move it. It obviously isn't working out the way he had planned for some reason. He may have decided he had to come here and take charge himself. I can't think of any other reason strong enough to bring him here."

"He won't come back to the castle again, will he?"

"No, Ann. There are policemen down in the courtyard now."

"Alex, Olga told me he was a friend of her husband's. That's why they let him in."

"I know. I tried to talk to von Toblen about him while the doctor was with you just now, but he refuses to say anything. He looks and acts like a guilty man, but he's trying to put a stone face on it. Why he is mixed up with Krueger I don't know yet. His wife apparently doesn't know anything about it. The only thing von Toblen would say to me was, 'I beg you not to discuss this matter with my wife until I have had a chance to speak with her.' Whatever it is, he wants to tell her himself."

"What about Karen? Has there been any sign of her?"

"No." He told her about the police finding the van. "They're working in the open now. The other members of the expedition have been told they aren't to leave Schloss Adler until further notice. They've been more or less confined to quarters. Now that it's known that Krueger is in the Tirol, he won't be allowed to get away this time. Even if the gold is never recovered, at least they'll get him."

"If you hadn't come here when you did you might not have found out he was here until it was too late. He might have finished what he came for and got away."

"That's very possible. It was only luck that Max

found out where you were when he did and let me know. Thank God he did." Then he said, "I really think you should rest now and not talk."

"I'm all right, honestly."

He sat with her for a while longer until she said, "Alex, what time is it?"

He looked at his watch. "Almost ten o'clock."

There was a lamp burning on the dresser. The curtains were still drawn. When she glanced toward the window he said, "In the evening." She had lost all track of time.

She sat up. "Alex, I'd really like to get out of here."

"Are you sure you're ready? Not too weak?"

"I'll be much better if I can just get away from this place. I don't want to spend another night here."

He helped her down the stairs and out to his car, past the somber policemen standing in the courtyard. They drove off, neither one looking back. They didn't talk until they were off the castle road. As they were crossing the bridge by the hayfield she said, "I hope they find Karen all right."

"So do I." He hadn't told her about the blood in the van. "The police are all over these mountains looking for her now. Don't worry about it. Are you feeling okay?"

"Better all the time."

"I'm taking you to the Bruners' and I'll stay with you until it's time for you to get some sleep. And after that—"

She remembered the last time at the Bruners'. "You'd better not."

He glanced at her in surprise. "Why not? I thought you liked them."

"I do. It's just that the last time I was there, the day we had the argument, Freida was there. Elsa said that Max thought she wasn't feeling too well, so he went to the *alm* and got her."

"I didn't know that. When was this?"

"You know, the day when we—" But it wasn't really an argument they had had. She had been the one who flew off the handle. "It was just yesterday when I was there, I guess. It seems a much longer time ago than that."

"No, I mean when did Max bring Freida home?"

She tried to remember. "The day before that, I think."

There was a brief silence. He reacted a few seconds later. "My God, why didn't I think of it?" He slowed the car and began turning around in the road.

She sat up. "Alex, what is it?"

"I know now where Karen Phillips is."

He parked down the road from the *alm* house. There was one lit window, and a kerosene lantern threw a pool of light across the middle of the bare porch. He left the keys in the car and got out.

"Roll up the windows and lock the doors," he told her. "If anything goes wrong—or even appears to go wrong—I want you to drive immediately back to Tellin. Can you do that?"

"Yes, but—"

"Go back to the castle and tell the police lieutenant there. All right?"

"Yes, but, Alex—Karen is here?"

"Max kidnaped her, I think."

"*Max?*"

"I can't explain now. Remember what I said." And he was gone. A few moments later she saw him step into the yellow circle of light on the porch and go in the front door.

EIGHTEEN

A flickering light moved past a window in the rear of the log house and then faded as it crossed into one of the opposite rooms. He must have found a lantern in the front room. She wished he had let her go with him. At least she would have known what was going on.

She couldn't hear anything outside with the windows rolled up. She moved over to the driver's side of the car and opened the window a couple of inches. What could be taking him so long? Either he couldn't find Karen or else he was talking with her. Or with Max. *Why would Max kidnap her?*

How long had he been gone? Ten minutes?

Fifteen? She realized she was probably overestimating the length of time.

As though someone had lit a match, the interior of the car grew bright. She whirled around. Headlights were moving up the road. The twin beams had already come much nearer before she thought to duck down against the seat. The bright headlights searched the interior of the car. She held her breath.

The driver didn't stop. After the vehicle passed she could tell by its taillights that it was an old pickup truck. It was parking farther down on the other side of the road, the lights off now. Someone got out. She leaned forward, keeping her head low and grasping the dashboard as she strained to make out who it was. She could only tell that it was a man. She wondered whether Max had an old pickup. She wasn't afraid of Max. How could anyone be afraid of Max?

The man walked past the house before he crossed the road. He stepped onto the far end of the porch and moved cautiously along the outer wall toward the door. She could see now that he was an older man with gray hair, but not thin enough to be Max Bruner. Even though she couldn't make out his face, she knew it was Heinrich Krueger. He didn't have on the dark suit he had worn as Dr. Praeger but was now dressed like a farmer. Just as he reached out to open the front door he lifted his other hand as well to waist level and she saw the clear silhouette of a pistol. He went inside.

Alex had told her to drive to Tellin and she had

intended to do as he asked. But the thought of turning the key in the ignition was impossible. By the time she got back, Alex would be dead. She fumbled in the glove compartment for some kind of weapon. There was nothing but a flashlight. She took it.

She got out and hurried up the dirt road, trying not to trip on the uneven ground. She had no idea what she could do when she got there, she only knew she couldn't do anything else. The flashlight was heavy. Maybe it would be heavy enough to hit Krueger with from behind if she got the chance. She *had* to.

The front door was open. She could see there was no one in the front room. She went inside. There were stairs on one wall next to a ceiling-high stove. On the other side, two rough wooden doors led off to the right. She stood still for a long while listening, trying to think. From behind the second door she heard the rumble of voices. She could make out Alex's voice, the words muffled. He sounded quite calm. A woman said something. That would be Karen. She didn't hear anyone else.

She crossed over quietly and listened again. Karen said something else and then there was silence for a moment.

Was Krueger in there with them? She couldn't be sure. But if he hadn't heard them talking, if he'd gone upstairs first, he might be down any minute. She had to warn Alex. She twisted the knob as quietly as possible and eased the door open.

Alex was standing across the room half facing

toward Karen. With an overturned chair behind her and a long piece of rope on the floor, Karen was rubbing her wrists and staring peevishly at the floor. Her long hair was uncombed. There was a red mark on her cheek, a wound days old. Ann opened the door a bit further, trying to see the rest of the room before she went in.

As soon as Alex saw her she knew from the anguished look on his face that she had made the wrong choice. Krueger stepped into her sight from behind the door and pointed the pistol at her in a casual way. He had, she noticed, shaved off the goatee. He bowed slightly, in an ironic way. "Please do join us, Miss Cole. I wondered if there might not be someone else in that car after all. I believe you know Herr Schuler. And this is my daughter, Maria Ortega y Perez. Whom you already know also, though by her American name, I believe." He was speaking in clear crisp German, and she had no difficulty at all in understanding him.

He came toward her and took the flashlight from her limp hand. She had forgotten to turn it off. He flicked the switch and motioned her to the other side of the room. She went. She couldn't look into Alex's eyes. She knew she had just thrown away his only chance.

Karen was Krueger's daughter. All at once she understood it. Krueger's daughter, born in South America but schooled in the United States. That hardness of hers was not after all a façade. She was the one who killed her sister. She was the one who sabotaged her car. The first time Ann had seen her she was working on a piece of machinery on the boat. She would know how to disconnect a brake line. She must have done it that evening after she had come to her room and found out that Ann was suspicious about Judy's death—and yes, she knew she had reacted too obviously to the mention of Argentina. She had given herself away. Karen must have destroyed her brakes sometime that night when the other divers were in the village, sometime before Max came to her room to lure her away from the castle. Max. Where did he fit into it?

But that couldn't matter now.

Karen saw Ann looking at her and glared back. "You should have stayed in New York. Why didn't you?" It was a shock to hear her speak German. They must have been a two-language family, her mother a South American woman, Krueger speaking his native language around the house wishing he were back in Germany. Yes, she could see that. And then Karen was sent away to school in America, so she learned English too. Her father must have wanted the best for her. Karen spat the question at her again. "Why didn't you stay home?"

"You know why."

Karen glanced away quickly. "Your sister should have minded her own business."

Krueger said sharply, "I want to know who this man is that brought Maria here, and where he is now."

"I told you I don't know," Alex said. How could he sound so calm, she wondered. She finally looked at him, and he put his arm around her shoulder.

Karen said, "I already told you, Papa. He left to get food for me. He said he wouldn't be back until morning."

Krueger flared up. "Keep quiet. You have botched everything and I haven't forgiven you. You should have *known* I would never be stupid enough to try to call you in Tellin, no matter what the circumstances."

The girl looked chastened, but she said, staring at the floor, "He told me the call came from Zurich, and I knew you were there. How could *he* have known where you were? I thought he had to be telling the truth."

He gave her a furious, withering look. "You have done nothing as you should have. And accomplished *nothing*."

"It wasn't so easy the way you told me it would be. And when I found the boxes, they were empty. He said he took the gold a long time ago. How is that my fault?"

But Krueger was staring now at Alex. "This man sent me a message in Zurich. 'I have your gold and your daughter. I will return both in exchange for information concerning the whereabouts of certain of your former associates.' "

So it was Max. Still searching for those men. Did he really take that gold? She watched Krueger, fascinated, while he said, "The man is obviously deranged. He may have underestimated me, thinking I would be too afraid to meet him. I am not. After I came to Tellin I received a second message. 'Your daughter is at the cabin on the highest *alm.*' Releasing her was to be a sign of his good faith, I suppose. Now he doesn't dare show himself, apparently. Unless it could be you, Herr Schuler? Or a partner of yours, perhaps?"

"I can't tell you anything, Krueger, because I don't know where the man is."

"We shall see."

"It isn't Alex," Ann said quickly. "Believe me—"

"Hush, Ann."

Krueger eyed them with calculation. Alex let his arm drop from Ann's shoulder. "Perhaps you are right," he said. "It would be better to have part of a million dollars than to die with nothing. Obviously you have the upper hand, Herr Krueger. I am willing to make a deal. My partner and I will split the gold with you."

Krueger studied him impassively. "On what terms?"

"I will take you to find him. We'll leave the women here. He's waiting for me in a farmhouse north of Tellin. He and I will show you where the gold is. We'll split it three ways." *No,* Ann thought. *Nice try, Alex. Even if it doesn't work, thank you.*

Krueger began to smile, though it wasn't really a smile. "You have a very inventive mind, Herr Schuler. Very admirable, but I don't believe you are telling me the truth. Though you do know something about this. You know how much gold there is."

"Of course. I weighed it myself."

Karen said, "He couldn't be in with the other

one, Papa. He came here and untied me. He was going to take me out of here, and he asked me where the other man was."

"Not a partner then," Krueger said to Alex. "A competitor perhaps. It doesn't matter, however. I don't have time to find out the truth. It's the other man I'm after. We'll go outside now." He didn't have to say what he was planning to do with them.

Karen led the way out the front door, in front of Alex. Krueger must have decided he could control Alex by following Ann out with the gun leveled at the small of her back. Karen stepped off the porch and marched away with Alex a few paces behind. Ann stepped down into the yard and glanced back. Krueger was still pointing the gun at her, and he was halfway across the porch. The lantern light threw his shadow across the door.

There was an ear-splitting gunshot. Ann's hands flew up to shield her face, and still she saw Krueger rise off his feet and jerk backward as though he had been pulled from behind by a rope around his waist. He slumped against the doorjamb, and a red hole began to grow large on his chest like a time-lapse film of a vibrant flower. Someone was screaming and she realized it was her own voice.

Alex grabbed her shoulders from behind and pushed. "Out of the light," he ordered. She staggered a few feet away and looked back to see him go to the man on the porch. He felt for a heartbeat in the neck, then let Krueger sag down to the floor. There was another scream now, a long wailing animal sound that came from Karen. She ran to the porch and knelt down to embrace her father, saying things in Spanish now, the soft language, but her father was past hearing them.

Alex went and stood on the edge of the porch, facing the direction from which the bullet had come. He didn't have to wait long.

Max appeared at the rim of the light's range holding a rifle under his armpit, muzzle down. Alex looked pale, but it was not from shock, Ann realized when he spoke. It was anger.

"I never knew you to be a damned fool, Max."

The other man looked very relaxed, just as though he were about to discuss their chess games or the relative toughness of the residents of the Tirol and New England. "A fool? Because I did what somebody should have done a long time ago?" He glanced at Krueger's body and the grieving girl on the porch. "I'm only sorry he didn't live to see me kill his murderess daughter."

Alex was already edging back down the porch, facing the other man all the while. "How do you know she's killed anyone?"

"Oh, I know. I wish I could have told you everything a long time ago, Alex. You would have really enjoyed this story. But I knew you couldn't have gone along with it. If you want to see a fool, though, look at that bastard on the porch. He was so sure of himself in 1945. Now look at him." He let out a short laugh, almost a bark, and Ann realized that she had been wrong to think he was calm—it was the false, deathly quiet of someone dangerously agitated.

"He thought," Max said, "that no one but him knew there was a cave under Schloss Adler. When I was a boy it was my favorite secret hiding place. A friend and I would crawl inside and stay there by the hour. We learned to tell the time of day it was just by the light filtering into the pool. We even took lines inside and fished right inside there. That's how we discovered for certain there was an opening into the lake, you see." He laughed again, remembering. "Some big fish took the bait and pulled the pole underwater before my friend could grab it. Later we found the pole floating outside in the lake. It would never occur to Krueger, you see, that any ordinary Tirolese could have known about the cave before *he* discovered it. Especially not two young boys.

"And that's not the only mistake he made, because he really *was* a fool. He didn't think the lowly Tirolese could form an underground. He didn't know I was there watching from outside my house that night. It was my assignment to watch and pass the word along if Krueger decided to take his troops out of there suddenly. My house was the perfect place for that job. But instead I saw a few of his men come down from the castle toward the lake. I was too far away to know what they were doing with those lights moving back and forth. But I found out soon enough. When I saw later that the entrance to the cave had been sealed and heard the rumors about the fortune in gold, I knew what he had been doing there."

"But you didn't report it to the Americans who were there looking for it." Alex had moved inconspicuously down the porch until he stood between Max and the girl, who seemed oblivious to both of them.

"Oh, I thought of it, I thought of it. But they said the gold had been melted down into ingots, you see, and there would be no way of finding the rightful owners. Everyone had heard that. So where would it go? What would they have done with it, the authorities? It would have gone into a national treasury to be squandered away or directly into someone's pockets. No, I didn't report it."

"You took the gold for yourself."

"Not one schilling." His voice was thick with indignation. "No. I waited until long after the Americans left before I bought a small diving tank and took it out with me on my boat. When no one was around I would slip overboard and search. At first I didn't take any of it out. I just opened the boxes and looked at it. Then I decided to use it to track down that—that dead animal over there." He raised the rifle to point and must have seen that Alex was in the line of fire. He didn't lower the gun.

"So you were the one who sent those anonymous notes to the police?"

"I was willing to give them a chance to catch him if they could. The detectives I hired, the ones who found him for me, told me he was having his daughter trained in diving. He was going to send *her*. Then I thought there might be a way to get him here too."

"So when Krueger sent his daughter here, you already knew who she was?" Alex was prompting him, Ann realized, to divert his attention, but what did he plan to do?

The older man nodded emphatically. "I knew her identity from the beginning. I used part of the gold to hire those detectives, you see. It took several years before they found someone, an old colleague of Krueger's who had accidentally found out where he was. For a price he gave us the information, and the detectives found Krueger in Argentina in 1965. I had them take a picture of him and send it to the authorities in Vienna, but nothing came of it. I was afraid that was the way it would be. They couldn't touch him as long as he stayed there. I had them continue to keep track of him for a long time, hoping he would come back. Nothing happened until the archaeologists discovered there was something of interest to them in the lake. I was certain Krueger would see that as an opportunity if he heard about it."

"Then you kidnaped Karen to force Krueger to come back to Austria. I understand. I only wonder why you waited so long after she got here."

"But it's simple, Alex. I wanted her to have a chance to send word to Krueger that the gold wasn't there any longer. I wanted to be sure he would believe me when I sent him the message that I had it. I didn't trust his fatherly impulses that far, you see. I wasn't sure he would risk coming here just for the girl."

"But how could you know when she had found out the gold wasn't there? How could you tell?"

"It was easy, my friend. I set a string not far inside the underwater entrance to the cave and attached it to a small float. Whenever I was out on the lake I would look for that float. When I found it one morning, I knew she'd been inside."

Maybe it was that day I saw him from the beach bringing in those nets, Ann thought. *Maybe he had picked up the float that morning.* Max said, "After she had been inside the cave and seen the empty boxes, she knew that she had killed that young girl for nothing."

"You knew it was Karen all along. You could have turned her over to the police, Max, and ended this long ago."

"But then I wouldn't have gotten Krueger. They wouldn't have either. Don't you understand, Alex?" His voice rose. "That's all I wanted. From the time when the lake site was discovered, I only prayed that he would hear about it and decide to take advantage of the situation. And he did. Only he didn't count on someone being smarter than he was. After I learned that he had sent for his daughter and brought her to Acapulco and that she was being trained in diving there, I was only afraid something might go wrong at the other end. Maybe Windle wouldn't get the money he needed, or maybe she wouldn't be accepted. So I fixed that too. I created Omega myself. I put almost all the rest of the money into it." He took a deep breath. He had finished what he had to say. "Step aside, Alex, and let me complete this job."

"Max, I can't do that."

His chin shot up in surprise. "She killed one girl and tried to kill your friend Ann. You can't be trying to protect that filth?"

"I'm trying to protect you, Max."

"Do you think I will let the authorities take her and let her go after a few years? Krueger's daughter?" He was coming closer to the porch, raising the rifle at Alex's chest as he came.

"Karen killed Judy Hamilton, yes. But if you were thinking straight, Max, you would see that if you hadn't set up the Omega Foundation that brought her and Karen here, Judy Hamilton would still be alive."

The old man's step faltered and he stopped. "No. No, Alex." His voice was imploring. He had thought about that before.

"You know I'm right. Karen might not have qualified for a regular expedition, but you saw to it that she got chosen to come here. You set up Omega so you could make damned sure Krueger got his way. You picked the others too, names pulled out of a hat as far as you were concerned, even the Latin Americans. What were they for, Max? To give the police some others to suspect while you played a game of revenge with Krueger? You didn't really *want* the police to discover who Karen was, did you?" He stepped down off the porch.

"Yes. No." He shook his head in exasperation. "Don't get me mixed up, now. This is something I have to do. For Freida's sake."

Karen was stirring behind them. She stood up

and looked at Max and at the rifle. Her face showed no fear, no apparent emotion of any kind.

Alex took another step toward his friend. The rifle muzzle was a little more than an arm's length away. "If you hadn't set up Omega with Krueger's gold, Karen wouldn't have killed anyone. You don't have any right to punish anyone, Max. Let me have that gun." He held out his hand. They stared at one another, neither man moving.

Max lowered his arms and let the rifle drop to the ground. Then he turned his back and walked away, putting his hands over his face.

"Alex, look out!" It was Ann shouting. She saw how Karen had taken that moment to pick up the pistol Krueger had sent scudding across the floor when he fell. She ran down the porch with it, jumping to the ground and running into the darkness at the side of the house.

Ann was hurrying toward Alex when another shot went off. It echoed from the wall of mountains above them. Then there was only silence as Alex gathered her into his arms and held her.

NINETEEN

They were up until dawn.

The policemen that Alex sent back to the *alm* found Karen's body near the back of the house. She had put a bullet in her temple. Max was inside, waiting for them. He came out when he heard the car drive up and told them he had killed the man on the porch. He told them who the man was, and then he surrendered peacefully.

Before Alex had left, Max asked him to tell Elsa what had happened, and so he and Ann stopped at the Bruner house on their way back. Ann went in with him. She thought Elsa might need someone. But Elsa took it quite stoically. She had already known enough to be prepared for the worst.

She sat with one rough hand on top of the other in her lap as she talked with them. She had known from the beginning that her husband had found the gold. "One day I saw the diving tank he had bought, and he told me. I tried to persuade him to turn the gold over to someone," she said. "But Max said it should be put to some good purpose, something appropriate to the memory of the people it had been taken from."

When the archaeological site was discovered and Max began to talk of how Krueger might finally come back she had begged him to leave matters alone, not to interfere, and he had promised to leave Krueger to the police. She had been afraid all along that he wouldn't. "Let the past remain in the past, is what I told him," she said.

"What will happen to Elsa now?" Ann asked him when they finally went back to the car.

"I don't know. She's a strong woman. She will survive this somehow." He sounded very tired.

"Without Max?"

"I don't think he'll be away for long. If the jury

knows about Freida. . . . Considering it was Krueger he killed, he may even get a medal."

Yet he didn't sound happy about it. "But it's good, isn't it," she said, "that he might not have to go to prison for a long time?"

"Prison might have been better for him in the long run than having people treat him as a hero. I know him, he is a good man. He has a conscience. I was too hard on him back there, but he'll never know whether he did the right thing."

"Because of my sister, you mean."

"That, and because Krueger and his daughter almost killed you too. It was only luck that they didn't."

"I can't blame Max for any of that."

"Neither do I. But he did make it possible for Krueger to set his plan in motion. Max can't evade that much of the responsibility. I can't evade my share of it, either."

She looked at him. His face was much older in the weak reflected light of the dashboard. "Don't say that."

"If we hadn't had that argument, which was really my fault, you wouldn't have gone back to Schloss Adler alone and Krueger wouldn't have had a chance to kill you." He pounded the steering wheel softly with his fist. "And besides *that:* If I had been using my head I would have realized there was only one reason for Karen to be kidnaped. And only one reason for Krueger to risk coming here. Karen was under suspicion, and I knew Max had good cause to hate Krueger. I should have realized sooner what was going on."

"Don't talk nonsense. There was no way for you to know that, Alex. I haven't forgotten that you saved my life, either. And if it hadn't been for you, Max would have shot Karen as well."

He didn't say any more. She knew he was still worried about what was going to happen to Max. They drove to his house and went inside. They sat at his kitchen table and tried to talk off the tension of the past few hours and sort out the remaining pieces of the puzzle.

They decided that Karen must have been concerned about how much Ann knew from the time she arrived at Schloss Adler because she didn't know how much Judy had told her in her letters. And they discussed the possibility that before he came to the castle for Karen Max may have learned which room was hers from the same cleaning woman who later told him that Ann was ill. Or he may have seen the three boys at the *Gasthaus* that night and followed them back, waiting until the lights went out in their rooms, as Martin had suspected. Then he took Karen up to the empty *alm* house, knowing the police wouldn't search there because of Freida.

"It's strange, but despite everything I can't help feeling sorry for Karen," Ann said. "With her father being who he was, what chance did she have? But it's pointless to think about all the *ifs*, isn't it?"

"It's the question of responsibility that I wonder about. Who is ultimately responsible for what Max did tonight? Max alone, or the soldiers who raped Freida? The German officers who should have restrained them, or punished them afterwards?"

"Or maybe Hitler for starting the war and bringing the soldiers here?"

Alex shrugged his shoulders wearily. "If you go back that far, you may as well include the ones who started the First World War, since the second was an outgrowth of that one, they say. Pretty soon we'll be saying it started in the Garden of Eden. No. It's the old saying, if everyone is responsible then no one is responsible."

"There's something else I don't understand though, Alex. Why weren't the police able to trace Karen back to Krueger?"

"Because he must have devised a false background for her way back when he decided to send her to America to be educated. He chose a name and an identity for her then and she had been building on it ever since. He must have felt she would stand a better chance if he made sure she could never be connected with him. After all, even in Argentina he must have been afraid someone would recognize him one day."

The archaeological group at Schloss Adler would have learned the news by now, they knew, from the policemen who were coming and going there. The expedition would probably be terminated—it would be interesting to know what Professor Windle would say when he found out his project had been financed not by an eccentric Latinophile millionaire but by a native of Tellin who was using Nazi gold he had taken out of a cave under the castle. Someone would have to tell

Windle the whole story about who Karen was and how she had murdered Judy.

They had been drinking coffee, and Ann had kept going for a long time on nervous energy as well as the caffeine, but when the windows began to grow light her weariness finally caught up with her. And she could tell by looking at him that he was exhausted too. He gave her the bedroom and he took the couch in the front room.

About noon she woke up. Her first thought was, *It's finally over.* Unable to go back to sleep, she got dressed and went downstairs. Alex had fallen asleep in his clothes, one arm flung over his eyes. The blanket had slipped to the floor. She picked it up and tucked it around him.

His face was stubbly with beard, but she loved the way he looked. He moved his arm and opened his eyes. "Sorry, I didn't mean to wake you," she whispered. "Go back to sleep."

He reached out to catch her arm. "You aren't getting ready to run off again, are you?" His voice was thick with sleep.

She smiled at him. She felt better than she had in a long time. "No. Just to the kitchen. I thought I'd start breakfast."

"Talk with me for a moment." He pulled her down to sit beside him. "What are you planning to do now?"

She dreaded thinking about that. "Oh, bacon and eggs."

"No. Idiot. Later on."

"I really should call my parents. I want them to know I'm all right. And tell them—"

"About Judy?"

"Yes, and all that's happened. But that part of it—I'd rather wait until I can tell them in person." She didn't want to think about leaving. She knew she loved Alex more than she had ever loved anyone. How could she leave him?

No, she told herself, the question was, how could she stay? She imagined herself settling down at the inn, coming down to his cottage every morning. He wasn't the kind of man who would take delight in being pursued by a woman, and that's what it would be. She could see herself coming down to the cottage to make his breakfast every day. Or maybe even moving into that room upstairs. Let's really be shameless about it, why don't we? He was in his thirties and not married, he had evidently decided marriage wasn't for him. He probably valued his freedom too much. As Judy would have put it, he was a true Sagittarius, a born bachelor. No matter when he was born. She didn't know. She didn't even know when his birthday was, for that matter.

It was surprising how many things she didn't know about him. Except everything that counted.

"After I call my parents," she told him, "I'll have to make arrangements to go home in a day or two." She had said it tentatively, hoping he would say he didn't want her to leave at all.

He said, "I wish you wouldn't go so soon."

So soon. Later, but not just yet. Time for some

fun together maybe, nothing permanent. If that was the way it was, what could she do about it?

He had put his arm around her. "I was hoping that before you went back, we could—"

She stood up abruptly. "I think I'd better start breakfast." She went into the kitchen. He got up and followed her and stood in the doorway, buttoning his shirt sleeves and studying her as she opened one cabinet door after another, trying to tell herself she was interested in finding a skillet. She wouldn't look at him.

"What's the matter, Ann?"

"I don't like to talk much before I've had coffee."

He smiled crookedly. "Well, that's good to know." She heard him go out of the room, heard the shower running.

Later he sat down at the table, clean-shaven, just as she was bringing the coffee over along with the scrambled eggs. Before they had a chance to begin eating there was a knock at the door and Alex got up. "Probably the police. For the last time, I hope."

But it wasn't the police. Ann recognized the voice in the other room and then she saw her, Olga von Toblen, dressed very formally in a dark dress with a white collar. She came in and began talking with Alex in an intense, small voice. Ann overheard part of it. "I beg you not to say . . . it will do no good now . . . in the past."

"You're not talking to the right person," Alex told her. "Ann was the one involved. It's her decision."

She had forgotten all about von Toblen's connection with Krueger.

The baroness glanced through the door and met her eyes, then came into the kitchen to stand uncertainly in front of her, not at all the imperious woman she had seemed when Ann first met her.

"Please sit down," Ann said, indicating the other chair.

"No. Thank you. I came to beg of you a favor." She pronounced the words carefully. She seemed to be making a special effort not to sound German.

"What is it?"

"You know, already, that my husband let Heinrich Krueger into our home, that he came to you posing as a doctor. Believe me, I did not know the man's identity. More important is that my husband did not know that he would try to harm you. Believe me, he didn't know that."

"He was supposed to be an old friend of your husband's, I thought."

"That wasn't true," she answered quickly. "Although I thought it was when I told you. Wilfred had never met Krueger before in his life, even though he knew well who he was. He did not even know Heinrich Krueger was still alive until he arrived in the middle of the night and forced him to open our home to him."

"If he knew who he was—"

"He had no choice." Her hand fluttered up to rub her eyebrows. She looked ill. "When you are my age you will realize that life is not as easy as we believe. My husband finally told me why he had to behave as he did. After we were put in prison in 1939 Wilfred agreed to—to co-operate with our enemies. They threatened to send me away, to kill me if he did not. He co-operated, and they let us go."

She was beginning to understand. "He collaborated with the Nazis?"

The baroness winced, but then she threw her shoulders back. "Yes, collaborated. It is a dirty word. People do in times of crisis things they believe they would never do. Yes. He informed on his fellow prisoners. Informed on the people he knew in the anti-Nazi movement. He has admitted it to me. After what happened to you, he realized he must tell me the truth. All these years I had thought it was the imprisonment that had changed him, all the suffering he had seen. But it was his own suffering." Her voice caught.

Ann felt sympathy for her, but she wasn't sure she could forgive her husband that quickly. "What do you want of me?"

"Only that you not reveal that my husband knew who Heinrich Krueger was when he came to us. Or that he threatened to reveal Wilfred's past. It isn't necessary for anyone to know."

She couldn't believe it. "You want me to lie for your husband, the way he's been lying all these years?" The people in Tellin, she knew, regarded him as a martyr.

"You must understand. Wilfred is ready to tell the police everything, but I have begged him not to. I told him it was for my sake, but I really do not care now. Except for him. The people here respect him. It would kill him if their attitude should change. His family name would be ruined forever. It would become a symbol of disgrace."

"And you aren't concerned about how that would affect you?" Ann asked skeptically.

"I know you believe me a vain woman, Miss Cole. I tell you sincerely I would not have come here to beg for myself. And had he known, my husband would not have let me come here for his sake. I only ask that you not mention it to the police that I told you this so-called Dr. Praeger was an old acquaintance of my husband's. I have told them that Krueger entered the house under false pretenses as a doctor for you, and that is all. I know they believe me, and they will not ask you about it, very likely. If only you will not bring it up to them."

"I don't know what to say to you." She was thinking of the lives Wilfred von Toblen must have ruined by working with the Nazis. She also remembered how sad and preoccupied he always looked. It couldn't be said that he had entirely escaped retribution.

Olga von Toblen spoke fervently. "Please do not say anything, I beg you." When she looked at her, Ann could tell that learning of her husband's past had already taken its toll. Wilfred von Toblen was an old man. Like Max, he didn't have many years left. There had already been so many ruined lives.

"All right. But if the police ask me the question directly, I won't lie."

The baroness made a quick movement to touch her hand in gratitude but drew back, fearing that it would offend her. "Thank you. Thank you," she said. She went out, and Ann heard her leave the house.

Alex had been waiting in the front room. When he came in she asked, "Do you think I did the right thing?"

"I don't know. But it's what I would have done."

They tried to finish the lukewarm eggs. As they were having a second cup of coffee he asked, "Are you ready to talk to me now?"

"What about?"

"Your plans."

She put her cup down. "All right."

"I know you have to go home, but do you suppose you could put it off for a few days? I'd like to take you back to Vienna with me, show you where I live, how I live."

A few days and good-by. Maybe it would be enough time to get him to change his mind about her. But what if it wasn't? "I don't know, Alex."

He could see that she wasn't willing to let him know what she was really thinking. He said, "Ann, I am too old to play games with you."

She wasn't going to play games either. "I really don't want to leave, Alex."

"And I don't want you to. But I know you have to talk to your parents. I hope you'll come back as soon as possible. That's why I'd like to take you to Vienna now. I think you would like it there. Like it, I hope, enough to come back to live there."

"Live in Vienna?"

"With me, I mean. I want you to marry me, of course."

Of course? "Alex, I—"

"And don't tell me you are still in love with what's-his-name. Because I wouldn't believe that."

She smiled at him. "I have forgotten all about what's-his-name."

"Well, then? Will you go to Vienna with me?"

"I don't have to see Vienna to know I want to marry you, Alex. But yes. I will. I thought—"

"What did you think?"

"That—" It seemed too ridiculous now to say it.

"Yes." He nodded gravely. He understood all right. A brief encounter, then good-by.

"But you see, that thought hasn't occurred to me since that first day I saw you." He looked at her and smiled, and then they both laughed. He reached across the table for her hand. "Come on, I'll go with you while you make your call."

It was going to be wonderful, she knew.